WEATHER AND WATER
RESOURCES

IMAGES, DATA, AND READINGS

DEVELOPED AT LAWRENCE HALL OF SCIENCE, UNIVERSITY OF CALIFORNIA AT BERKELEY
PUBLISHED AND DISTRIBUTED BY DELTA EDUCATION

FOSS Middle School Project Staff and Associates

FOSS Middle School Curriculum Development Team

Linda De Lucchi, Larry Malone, Co-directors; **Dr. Lawrence F. Lowery,** Principal Investigator
Susan Kaschner Jagoda, Denise Soderlund, Dr. Jan Woerner, Dr. Susan Brady,
Teri Dannenberg, Dr. Terry Shaw, Curriculum Developers
Dr. Kathy Long, Assessment Coordinator
Carol Sevilla, Graphic Artist; **Rose Craig,** Artist
Alev Burton, Administrative Support; **Mark Warren,** Equipment Manager

ScienceVIEW Multimedia Design Team

Marco Molinaro, Director
Leigh Anne McConnaughey, Producer and Principal Illustrator; **Rebecca Shapley,** Revisions Producer
Guillaume Brasseur, Producer and System Administrator
Dan Bluestein, Lead Programmer and System Administrator; **Roger Vang,** Programmer
Jerrold Connors, Senior Illustrator; **Bonnie Borucki, Sue Whitmore,** Illustrators
Alicia Nieves, Quality Assurance; **Coe Leta Finke,** Usability Review

Special Contributors

Marshall Montgomery, Materials Design; **John Quick,** Video Production; **Ronald Holle,** Content Consultant
John Jensenius, Warning Coordination Meteorologist, National Weather Service, Gray, MA
Warren Blier, Science and Operations Officer, National Weather Service Forecast Office, Monterey, CA
Patty A. Watts, Eric A. Pani, Water-Cycle Game
Learning about the hydrologic cycle and global climate change: A demonstration. Preprint
Fifth International Conference on School and Popular Meteorological and
Oceanographic Education: Weather, Ocean, Climate
Australian Meteorological and Oceanographic Society, Melbourne, 5–9 July 1999, 206–209

Additional Credits

Atmospheric Sciences, University of Illinois at Urbana-Champaign, WW2010 Project
Environmental Protection Agency; National Aeronautics and Space Administration
National Oceanographic and Atmospheric Administration; NOAA Climate Prediction Center
National Park Service, Johnstown Flood National Memorial
National Severe Storms Laboratory; National Weather Service

Delta Education FOSS Middle School Team

Bonnie Piotrowski, FOSS Managing Editor; **Mathew Bacon, Grant Gardner, Tom Guetling, Joann Hoy,**
Dana Koch, Cathrine Monson, John Prescott, Rebecca Waites

National Trial Teachers

Karen Burton, Cooper Middle School, Fresno, CA; **Linda Stewart,** Ahwahnee Middle School, Fresno, CA
Ted Stoeckley, Hall Middle School, Larkspur, CA
Laurie Erskine-Farley and **Peter Josefsson,** Parsons Middle School, Redding, CA
Lisa Evans, Southern Oaks Middle School, Port St. Lucie, FL
Gayle Dunlap and **Donna Moran,** Walter Bergen Middle School, Bloomingdale, NJ
Joan Caroselli, J. E. Soehl Middle School, Linden, NJ; **John Kuzma,** McManus Middle School, Linden, NJ
Terry Shaw and **Joe Green,** Irving Middle School, Norman, OK
Venus Ludovici, Janet McKenna, and **Cheryle Jackson,** Jay Cooke Middle School, Philadelphia, PA
Melissa Gibbons and **Doris Taylor,** Dunbar Middle School, Fort Worth, TX
Bebe Manning, Redwater ISD, Redwater, TX; **Belinda Simpson,** Redwater Middle School, Redwater, TX

FOSS for Middle School Project
Lawrence Hall of Science, University of California
Berkeley, CA 94720 510-642-8941

Delta Education
P.O. Box 3000 80 Northwest Blvd.
Nashua, NH 03063 1-800-258-1302

The FOSS Middle School Program was developed in part with the support of the National Science Foundation Grant ESI 9553600. However, any opinions, findings, conclusions, statements, and recommendations expressed herein are those of the authors and do not necessarily reflect the views of the NSF.

542-1437
Weather and Water
ISBN-10: 1-58356-432-2
10 11 12 13 14 15 QUE 13 12 11 10 09 08
ISBN-13: 978-1-58356-432-5

RESOURCES BOOK

Table of Contents

READINGS

DATA

READINGS
Table of Contents

Naming Hurricanes

September 15, 1999—Floyd threatens the eastern coast of Florida today with waves over 15 meters high and winds over 250 kilometers an hour. Meteorologists project that Floyd may continue its northward track as far as the North Carolina/Virginia border. Its effects will be widespread.

Who is Floyd? And who are Camille, Hugo, Agnes, and Andrew? They're hurricanes! Throughout human history hurricanes have caused destruction. Some of the most powerful hurricanes to come on land in the United States were those named above. How did these storms get names and why?

Because they have reputations as troublemakers, hurricanes are under the watchful eye of meteorologists from the time they first take form as tropical storms. In order to refer to them over a period of days and weeks, they are given names. It's much easier to talk about what Gertie did today than "that hurricane 18°N, 55°W." It also helps reduce confusion when more than one hurricane is brewing in the same area.

Naming hurricanes has undergone a number of changes over the years. In the West Indies (perhaps for reasons of protection), hurricanes were first named after the saint whose day fell closest to the day of arrival of the hurricane. Early in the 20th century an Australian forecaster started naming hurricanes for politicians. People thought this was funny because news reports could include headlines like "Dundee causing great distress," or "Gibson wandering aimlessly, but could be dangerous."

For a few years, the international phonetic alphabet (Able, Baker, Charlie...) was used to name hurricanes. The first hurricane of the season was Able, Baker was second, and so on. During World War II, U.S. Air Force and Navy meteorologists named the Pacific storms for their wives or girlfriends. Starting in 1953 the United States used whatever names they liked...as long as they were women's names, and the first hurricane of the season had a name starting with A.

In 1977 the World Meteorological Organization, a United Nations group, decided on the official naming system used today. They came up with 6 years' worth of names for storms in the Atlantic Ocean. After 6 years, they start over at the beginning. The six lists are alphabetical, and the letters Q, U, X, Y, and Z aren't used. It hasn't happened yet, but if all 21 names on a year's list are used up, additional storms would be assigned

Greek letters, like Alpha and Beta. The first forecaster who came up with the 21 names used a baby-naming book. He added names of his relatives to the list, too.

The first lists contained only female names. In 1978, the lists were changed to include both male and female names. They also include French and Spanish names for Atlantic storms.

Fourteen countries of the western North Pacific and South China Sea have approved new lists of names for tropical storms in the South Pacific. The names derive from many languages, and include Nakri (a type of flower in Cambodia), Haishen (a Chinese sea god), Pabuk (a big freshwater fish in Laos), Parma (a dish of ham, liver, and mushrooms popular in Macao), Ewinlar (a Chuuk storm god in Micronesia), Cimaron (a Philippine wild ox), Durian (a Thai fruit), and Conson (a mountain in northern Vietnam).

The names of the most severe storms are taken off the list for at least 10 years. At this time the exiled names include Fran, Andrew, and Hugo. The country that suffered the most destruction from the hurricane gets to add a new name.

When do storms get a name? Hurricanes and tropical storms are named once they start rotating and reach a wind speed above 65 kilometers (km) per hour (39 miles per hour). A tropical storm becomes a hurricane when it reaches speeds of 123 km per hour (74 miles per hour) or more.

Storm Names for 2002–2007 (Atlantic Storms)

2002	2003	2004	2005	2006	2007
Arthur	Ana	Alex	Arlene	Alberto	Andrea
Bertha	Bill	Bonnie	Bret	Beryl	Barry
Cristobal	Claudette	Charley	Cindy	Chris	Chantal
Dolly	Danny	Danielle	Dennis	Debby	Dean
Edouard	Erika	Earl	Emily	Ernesto	Erin
Fay	Fabian	Frances	Franklin	Florence	Felix
Gustav	Grace	Gaston	Gert	Gordon	Gabrielle
Hanna	Henri	Hermine	Harvey	Helene	Humberto
Isidore	Isabel	Ivan	Irene	Isaac	Ingrid
Josephine	Juan	Jeanne	Jose	Joyce	Jerry
Kyle	Kate	Karl	Katrina	Kirk	Karen
Lili	Larry	Lisa	Lee	Leslie	Lorenzo
Marco	Mindy	Matthew	Maria	Michael	Melissa
Nana	Nicholas	Nicole	Nate	Nadine	Noel
Omar	Odette	Otto	Ophelia	Oscar	Olga
Paloma	Peter	Paula	Philippe	Patty	Pablo
Rene	Rose	Richard	Rita	Rafael	Rebekah
Sally	Sam	Shary	Stan	Sandy	Sebastien
Teddy	Teresa	Tomas	Tammy	Tony	Tanya
Vicky	Victor	Virginie	Vince	Valerie	Van
Wilfred	Wanda	Walter	Wilma	William	Wendy

Weather Tools

Thermometer

A thermometer measures temperature. Temperature is measured in degrees. This thermometer uses the Celsius scale (°C). You would read the temperature as 21°C in the example above.

Some thermometers use the Fahrenheit scale (°F).

Compass

A compass is used to determine the direction from which the wind is coming.

- Stand facing the wind.
- Hold the compass in front of you.
- Rotate the compass body until the N lines up with the needle.
- A line running from the center of the compass needle straight into the wind tells you the wind direction.

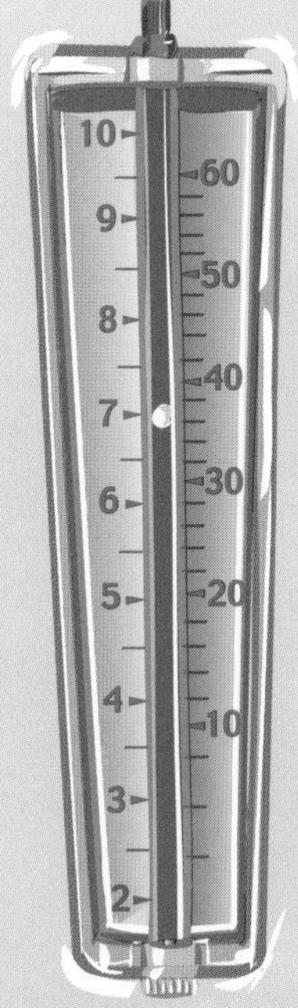

Anemometer

An anemometer measures wind speed.

- Face into the wind.
- Hold the meter in front of you straight up and down, scale toward you. Don't cover the holes at the bottom.
- The height of the ball indicates the wind speed. This picture shows 7 miles per hour.

If the wind is strong, cover the top hole with your finger to read the high scale on the right.

Hygrometer

A hygrometer measures relative humidity. Relative humidity is a measure of the amount of water vapor in the air. It is a percentage. On the hygrometer above, relative humidity is 63%.

Barometer

A barometer measures air pressure. Air pressure on this barometer is measured in millibars (mb). The barometer above indicates a pressure of 985 mb.

Air? Can't see it. Can't taste it. Can't smell it. If you pay attention, you might feel it as a gentle breeze brushing across your skin. Because we are so insensitive to air, it is difficult to understand what it is. Is it one thing, or a mixture of things? And where is it? Is it everywhere or just in some places?

As we go about our everyday business, we usually travel with our feet on the solid Earth and our heads in the atmosphere. The atmosphere completely surrounds us, pressing firmly on every square centimeter of our bodies—top, front, back, and sides. Even if we attempt to get out of the atmosphere by locking ourselves inside a car or hiding in a basement, the atmosphere is there, filling every space we enter.

An **atmosphere** is the layer of gases that surrounds a planet or star. All planets and stars have an atmosphere around them. The Sun's atmosphere is hydrogen. Mars has a thin atmosphere of carbon dioxide with a bit of nitrogen and a trace of water vapor. Mercury has almost no atmosphere at all. Each planet is surrounded by its own mixture of gases.

Earth's atmosphere is composed of a mixture of gases we call **air.** Air is mostly nitrogen (78%) and oxygen (21%), with some argon (0.93%), carbon dioxide (0.03%), ozone, water vapor, and other gases (less than 0.04% together).

Nitrogen (N_2) is the most abundant gas in our atmosphere. It is a stable gas, which means it doesn't react easily with other substances. When we breathe air, the nitrogen goes into our lungs and then back out unchanged. We don't need nitrogen gas to survive.

Oxygen (O_2) is the second most abundant gas. It takes up about 21% of the air's volume, and, because the oxygen atom is larger than the nitrogen atom, it accounts for 23% of air's mass. Oxygen is a colorless, odorless, tasteless gas. It is the most plentiful element in the rocks of Earth's crust. Oxygen combines with hydrogen to form water. Without oxygen, life as we know it would cease to exist on Earth.

Oxygen and nitrogen are called **permanent gases.** The amount of oxygen and nitrogen in the atmosphere stays constant. The other gases in this chart are also permanent gases, but are found in much smaller quantities.

Permanent Gases of the Atmosphere

Gas	Percentage by volume
Nitrogen	78.08
Oxygen	20.95
Argon	0.93
Neon	0.002
Helium	0.0005
Krypton	0.0001
Hydrogen	0.00005
Xenon	0.000009

Air also contains **variable gases.** The amount of a variable gas changes in response to activities in the environment.

Water vapor (H_2O) is the most abundant variable gas. It makes up about 0.25% of the atmosphere's mass. The amount of water vapor in the atmosphere changes constantly. Water cycles between Earth's surface and the atmosphere through evaporation, condensation, and precipitation. You can get a feeling for the changes in atmospheric water vapor by observing clouds and noting the stickiness you feel on humid days.

Carbon dioxide (CO_2) is another important variable gas. It makes up only about 0.036% of the atmosphere. You can't see or feel changes in the amount of carbon dioxide in the atmosphere.

Carbon dioxide plays an important role in the lives of plants and algae. Carbon dioxide is removed from the air during **photosynthesis.** Plants and algae convert light energy into chemical energy by making sugar (food) out of carbon dioxide and water. In the process, oxygen is released to the atmosphere. When living organisms use the energy in food to stay alive, oxygen is removed from the air and carbon dioxide is returned to the air.

Variable Gases of the Atmosphere

Gas	Percentage by volume
Water vapor	~ 0.25
Carbon dioxide	~ 0.036
Ozone	~ 0.01

There are other gases that you may have heard about. **Ozone (O_3)** is a variable gas. It is a form of oxygen that forms a thin layer in the stratosphere. Ozone is absolutely essential to life on Earth because it absorbs deadly ultraviolet radiation from the Sun. But ozone in high concentration can cause lung damage. In the lower atmosphere, ozone is an air pollutant.

Methane (CH_4) is a variable gas that is increasing in concentration in the atmosphere. Scientists are trying to figure out why this is happening. They suspect several things. Cattle produce methane in their digestive processes. Methane also comes from coal mines, oil wells, and gas pipelines, and is a by-product of rice cultivation. Methane absorbs heat coming up from Earth's surface.

These gases are all mixed together, so that any sample of air is a mixture of all of them. If you rise higher in the atmosphere, there are fewer molecules, but the ratio of each gas to the other is the same. The mixing is caused by the constant movement of the air in the part of the atmosphere near Earth's surface. Above about 90 kilometers, there is much less mixing. Very light gases (hydrogen and helium, in particular) are more abundant above that level.

Think Questions

1. **What is the difference between permanent gases and variable gases in the atmosphere?**
2. **During the daylight hours, plants and algae take in carbon dioxide and release oxygen. If humans continue to destroy rain forests, what might happen to the balance between these gases?**

A Thin Blue Veil

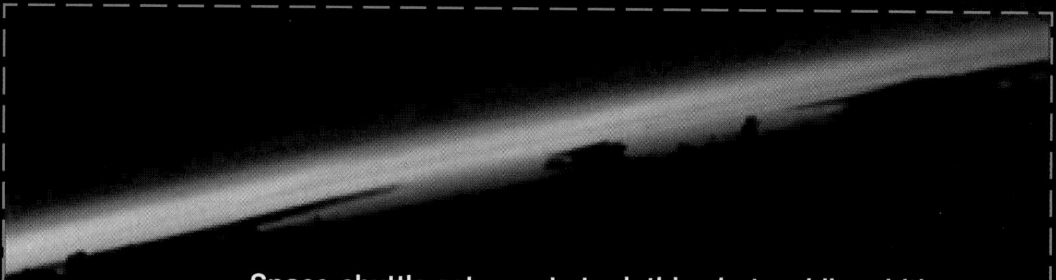

Space-shuttle astronauts took this photo while orbiting Earth. You can see a side view of Earth's atmosphere. The black bumps pushing into the troposphere are tall cumulus clouds.

The crew of Apollo 17 took this photograph of Earth in December 1972, while on their way to the Moon. The small box at the top of this image shows an area equal to the atmosphere image above taken by the space-shuttle astronauts.

It is cold in deep space. The temperature is in the neighborhood of –270°C. That's nearly 200°C colder than it has ever been on Earth. Near stars, like the Sun, it's hot—outlandishly hot—reaching thousands of degrees. There are, however, a few places here and there in the universe where the temperature is between the extremes. Earth is one of those places. In fact, the average temperature on Earth is just about the temperature of Baby Bear's porridge—not too hot and not too cold, but just right.

On a typical day, the temperature range on our planet is only about 100°C, from maybe 45°C in the hottest place to –55°C at one of the poles. The measured extremes are 58°C in El Azizia, Libya, recorded on September 13, 1922, and –89°C in Vostok, Antarctica, on July 21, 1983. That's a range of temperature on Earth of 147°C.

It's not only because we are at the right distance from the Sun that Earth has tolerable temperatures. Earth is wrapped in a blanket of gases—the atmosphere. Earth's atmosphere keeps the temperature within a narrow range that is suitable for life.

From space, Earth's atmosphere looks like a thin blue veil. Some people like to think of the atmosphere as an ocean of air covering Earth. The depth of this "ocean" is about 600 kilometers (km). The atmosphere is densest right at the bottom where it rests on Earth's surface. It gets thinner and thinner (less dense) as you move away from Earth's surface. There is no real boundary between the atmosphere and space. The air just gets thinner and thinner until it disappears.

Imagine a column of air that starts on Earth's surface and extends up 600 km to the top of the atmosphere. Scientists have discovered several distinct layers in this column of air. Each layer has a different temperature. Here's how it stacks up.

The layer we live in is the **troposphere.** It starts at Earth's surface and extends upward for 9–20 km. Its thickness depends on the season and where you are on Earth. Over the warm equator, the troposphere is a little thicker than it is over the polar regions, where the air is colder. It also thickens during the summer and thins during the winter. A good average thickness for the troposphere is 10 km.

This ground-floor layer contains most of the organisms, dust, water vapor, and clouds found in the entire atmosphere. For that matter, it contains most of the air as well. And, most important, weather happens in the troposphere. The troposphere is where the action is. This is where differences in air temperature, humidity (moisture), pressure, and wind occur.

These properties of temperature, humidity, pressure, and wind are called **weather factors.** Meteorologists launch weather balloons twice each day to monitor weather factors. The balloons float up through the troposphere to about 18 km. Weather factors will be investigated in detail as we continue to study weather.

The troposphere is the thinnest layer—only about 2% of the depth of the atmosphere. It is the densest layer, however, containing four-fifths (80%) of the total mass of the atmosphere.

Earth's surface (land and water) absorbs heat from the Sun and warms the air above

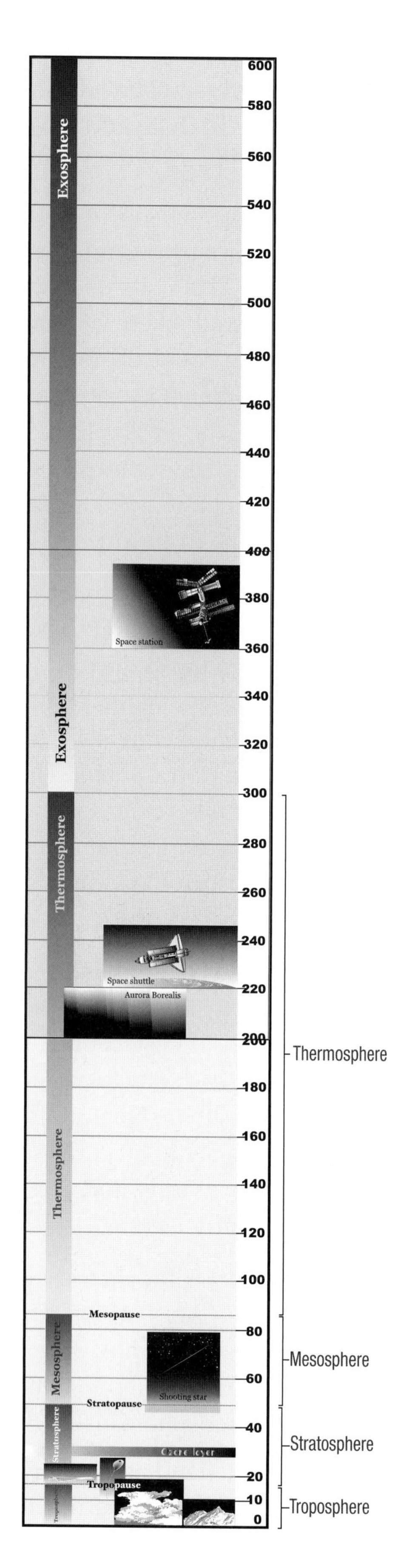

it. Because air in the troposphere is heated mostly by Earth's surface, the air is warmest close to the ground. The air temperature drops as you go higher. At its upper limit, the temperature of the troposphere is about –60°C. The average temperature of the troposphere is about 25°C.

Mount Everest, located in Nepal and Tibet, is the highest landform on Earth, rising 8.848 km into the troposphere. The air temperature at the top of the mountain is well below freezing most of the time. There is also less air to breathe at the top of Mount Everest. Climbers usually bring oxygen along to help them survive the thin air.

The **stratosphere** is the layer above the troposphere. It is 10–50 km above Earth's surface and contains almost no moisture or dust. It does, however, contain a layer of ozone (O_3), a form of oxygen, that absorbs high-energy ultraviolet (UV) radiation from the Sun. The temperature stays cold until you reach the upper reaches of the stratosphere, where energy absorption by ozone warms the air to about 0°C.

The jet stream, a fast-flowing river of wind, travels generally west to east in the region between the lower stratosphere and the upper troposphere. Many military and commercial jet aircraft take advantage of the jet stream when flying from west to east.

The **mesosphere** is above the stratosphere, 50–80 km above Earth's surface. The temperature plunges again, reaching its coldest temperature of around –90°C in the upper mesosphere. This is the layer in which meteors burn up while entering Earth's atmosphere, producing what we call shooting stars.

Beyond the mesosphere, 80–300 km above Earth, is the **thermosphere.** The thermosphere is the least-understood layer of the atmosphere and the most difficult to measure. The air is extremely thin. The thermosphere is the region of the atmosphere that is first heated by the Sun. A small amount of energy coming from the Sun can result in a large temperature change. When the Sun is extra active with sunspots or flares, the temperature of the thermosphere can surge up to 1500°C or higher!

Within the thermosphere is a layer noted for its chemistry, the **ionosphere.** The ionosphere contains a large number of electrically charged ions. Ions form when intense radiation from the Sun hits atoms and molecules. The ionosphere is responsible for the aurora borealis, or northern lights, and the aurora australis, or southern lights.

The identification of these four layers is based on temperature. There are no sharp boundaries or abrupt changes in gas composition between them. As average temperatures change with the seasons, the boundaries between layers may move up or down a little.

Beyond the thermosphere, Earth's atmosphere makes a transition into space. This area is the **exosphere** where atoms and molecules escape into space. It extends from 300 to 600 km above Earth. In this region, the temperature plunges to the extreme –270°C of outer space, and the concentration of atmospheric gases fades to nothing.

That 600-km column of air pushes down on the surface of Earth with a lot of force. We call the force **air pressure,** or **atmospheric pressure.** We are not aware of it because we are adapted to live under all that pressure, but there is a mass of about 1 kilogram (kg) pushing down on every square centimeter of surface on Earth. Your head has a surface area of about 150 cm^2. This means you have about 150 kg of air parked on your head. That's about like having a kitchen stove or a motorcycle pushing down on your head all the time!

Another way to look at it...if all the air were replaced with solid gold, the entire planet would be covered by a layer of gold a little more than half a meter deep. The mass of the entire atmosphere is about equal to half a meter of gold, but the atmosphere is much more valuable.

Think Questions

1. How is Earth's atmosphere like an ocean? How is it unlike an ocean?
2. Why do you think airplanes don't fly high in the stratosphere?

Wendy and Her Worldwide Weather Watchers

Wendy loves weather. She watches weather come across the Potomac River outside her kitchen window while she eats breakfast and watches weather forecasts on the TV. On her bookcase is a weather atlas, and she follows reports of severe weather on the World Wide Web. Wendy is planning to be a meteorologist.

It's no accident that weather is of interest to Wendy—she's seen a lot of it. Her dad is an agricultural consultant for the government, so Wendy's family has traveled around the world. She's been rained on in the rain forest, baked in the desert, blown over on the open plains, and nearly frozen in the tundra.

As a result of her travels, Wendy has made a lot of friends around the world. And she keeps in touch with them by e-mail all the time. They have formed a little weather study group and often share information about the weather in their local areas. They like to call themselves the Worldwide Weather Watchers. Some people put a finger in the air to figure out which way the wind is blowing; Wendy puts a finger on her keyboard.

On June 20, Wendy listened to a TV weather report. The meteorologist announced that the summer solstice would be tomorrow, on June 21. He said it would be the longest day of the year. You can count on more daylight on the solstice than you will see again for a whole year.

"Summer solstice...longest day of the year...I need to find out more about this," Wendy thought. "I think I need some help from the Worldwide Weather Watchers." She went to her computer and sent out a message to all of her friends. It said,

Here in Virginia it is the summer solstice on June 21. It is going to be the longest day of the year. What I want to know is if it is going to be summer solstice in your town, and if it is the longest day of the year for you, too. I'm also wondering, if it is the longest day of your year, how long is it?

What I would like you to do is find out when sunrise and sunset are in your town. If you give me those two pieces of information, I can figure out the rest. I plan to be up the whole day. I'm setting my clock to get up early so I won't miss a minute of the longest day of the year.

Thanks for helping me with this project. Wendy

Before going to sleep that night, Wendy rechecked her clock radio to make sure it was set for 5:30 a.m. As she pushed her clock back into place, her computer beeped. It was her first report—her friend Shawn from Auckland, New Zealand. The message didn't seem right, though. Shawn reported that the Sun rose at 7:48 a.m. and would set at 4:55 p.m.! That was just over 9 hours of daylight. How could that be the longest day of the year? Had she heard the TV meteorologist right? Had Shawn reported his numbers incorrectly? Wendy climbed into bed, pulled the covers up around her chin, and let the questions turn over in her mind as she drifted off to sleep.

Wendy awoke to "a 40% chance of thunder showers by late afternoon, clearing by nightfall. Winds from the southeast at 10 knots...." She bounced out of bed and pulled back the curtain. Dark! Yes, up before the Sun. She checked her e-mail and found three more reports from the Weather Watchers: Hiroko from Sendai, Japan; Seeta in New Delhi, India; and Makindu in Nairobi, Kenya.

Wendy hurried down to the kitchen for breakfast. The kitchen window had the best view to the east. Already the darkness on the horizon was yielding to the first suggestion of sunrise. By the time Wendy took her first bite of toast, a line of orange had pushed between the horizon and the darkness above. The kitchen clock showed 5:40 a.m. Sunrise had to be pretty soon. Wendy peeled an orange. The sunrise intensified and color moved across the bottoms of the clouds. She could see the brightest place on the horizon clearly now. Night had been replaced by the gray of early morning. Then the very tip-top of the red orange Sun peeked over the horizon. Wendy glanced at the clock. It was 5:44 a.m.—sunrise on the summer solstice.

Wendy watched as the complete disk of the Sun glided free of the horizon and hung suspended in the sky. Then she returned to her bedroom to check her computer for more messages. Justin from Punta Arenas, Chile; Maria from Quito, Ecuador; Billy from Barrow, Alaska; and Elke in Stockholm, Sweden. She read through the reports quickly, but the information was all a blur. There didn't seem to be a pattern to the data. How could she make sense of them?

She wrote down the sunrise and sunset times and returned to the kitchen. Her dad was opening the paper and eating yogurt. "Dad, I got up to see the sunrise because this is the longest day of the year. I'm going to see all of it. The Sun first appeared at 5:44, and I plan to see it go down, too."

"Good for you. Where will you go to watch the sunset? Our apartment has a great sunrise view, but no sunset view."

Wendy gulped. She hadn't thought of that. "I don't know," she admitted.

"How about this for a plan. I'll make sure I'm home in time for us to go up on top of the building to see old Sol set. What time will the Sun be setting?"

Again, Wendy was caught off guard. Her dad tossed her the paper and said, "Look it up. I need to gather a few papers before I head off. Give me the time on my way out."

"Of course," breathed Wendy, "the weather page." She dove into the paper, threw it open to the weather page, and quickly found the information she wanted. June 21, summer solstice, sunrise at 5:44 a.m., sunset at 8:37 p.m.

"Sunset is at 8:37 p.m.," Wendy called up to her dad, "but could you be here a little early? I don't want to miss this."

"I'll be here in plenty of time."

Wendy returned her attention to the e-mail data. She decided to organize the numbers in a chart. After thinking about it for a while, she put the cities in order from east to west, starting with New Zealand. This is the chart she produced.

City and country	Sunrise	Sunset	Length of day
Auckland, New Zealand	7:48	4:55	9:07
Sendai, Japan	4:18	7:06	14:48
New Delhi, India	5:22	7:18	13:56
Nairobi, Kenya	6:34	6:34	12:00
Stockholm, Sweden	2:48	8:59	18:11
Punta Arenas, Chile	8:00	3:32	7:32
Alexandria, VA, USA	5:44	8:37	14:53
Quito, Ecuador	6:22	6:22	12:00
Barrow, AK, USA	None	None	24:00

Wendy studied the chart. The hours of daylight varied widely, from less than 8 hours to 24 hours. More information would be needed to make sense of the data. She went to her weather atlas, opened to the map of the world, and located the countries in which her Worldwide Weather Watchers lived. She particularly wanted to see what might account for the huge difference in daylight between Justin's home in Chile and Billy's in Alaska.

"Hmmm," Wendy thought, "Justin lives way down in the southern tip of South America, and Billy is close to the North Pole in Alaska. I think I need to add latitude to my chart and put the cities in order from northernmost location to southernmost location." When Wendy added latitude to her chart and reorganized, this is what she saw.

June 21

City and country	Latitude	Sunrise	Sunset	Length of day
Barrow, AK, USA	71°N	None	None	24:00
Stockholm, Sweden	59°N	2:48	8:59	18:11
Sendai, Japan	38°N	4:18	7:06	14:48
Alexandria, VA, USA	38°N	5:44	8:37	14:53
New Delhi, India	28°N	5:22	7:18	13:56
Quito, Ecuador	0°	6:22	6:22	12:00
Nairobi, Kenya	1°S	6:34	6:34	12:00
Auckland, New Zealand	37°S	7:48	4:55	9:07
Punta Arenas, Chile	53°S	8:00	3:32	7:32

One thing became clear to Wendy. Latitude did relate to the length of the day. But why did locations in the northern latitudes have longer days than locations in the southern latitudes? And did northern locations always have longer days, or were days longer in the south at a different time of the year? More data would be needed to answer these new questions.

The U.S. Naval Observatory maintains lots of information related to Earth motions in the Solar System. Wendy had used their website in the past to check on the phases of the Moon. She thought she might find sunrise and sunset data there. To her delight, she could call up the sunrise and sunset data for any location on Earth for any day she chose. What a great resource! But what should she look up?

Some little voice inside told her to check the sunrise and sunset data for a date exactly half a year earlier. It took almost an hour to get the data, but it was worth the effort. This is the table she produced for sunrise and sunset on the winter solstice, December 21, for the previous year.

"Very interesting!" Wendy commented to herself. "The locations that have the longest days today had the shortest days half a year ago. And vice versa; the places with short days today had long days half a year ago. But look at Nairobi and Quito. Their days stayed the same all the time—equal amounts of daylight and darkness."

It *was* interesting. Wendy thought about day length all afternoon. She discovered that, if she added together the summer solstice day lengths for her town, Alexandria, and Shawn's town, Auckland, the sum was just about 24 hours. Alexandria and Auckland are almost the same latitude, but one is north and the other south. She also found that, when she added together the length of the day on the summer solstice and the winter solstice for any city, the sum was very close to 24 hours. What did it all mean?

December 21

City and country	Latitude	Sunrise	Sunset	Length of day
Barrow, AK, USA	71°N	None	None	00:00
Stockholm, Sweden	59°N	9:53	16:03	6:10
Sendai, Japan	38°N	7:00	4:32	9:32
Alexandria, VA, USA	38°N	7:23	4:50	9:27
New Delhi, India	28°N	7:03	5:31	10:28
Quito, Ecuador	0°	6:14	6:22	12:08
Nairobi, Kenya	1°S	6:24	6:35	12:11
Auckland, New Zealand	37°S	4:56	7:37	14:41
Punta Arenas, Chile	53°S	3:50	8:46	16:56

Right on the stroke of 8:00 p.m., Wendy's dad came through the door. "What a day! This feels like the longest day of my life. I'm bushed."

"It *is* the longest day...at least of the year...at least for us. Did you know this is the shortest day of the year for Shawn and Justin?" asked Wendy. Her dad raised his eyebrows. "Come on, Dad, let's go up on the roof. It's been a busy day for me, too, but it isn't over yet. I still need to see the curtain go down on the year's longest day."

From the roof, they had an unrestricted view in all directions. They could look north across the river to the nation's capital, Washington, DC, and west across rolling woods. The Sun peeked through spaces in the quickly dispersing clouds. The day was nearing sunset.

"Dad, I've been thinking all day about the different lengths of daylight that are happening around the world today. Billy says the Sun has been up since the middle of May and won't set at all until August. Makindu reports nothing surprising about today, 12 hours of daylight followed by 12 hours of darkness, just like every other day. Justin gets only 8 hours of sunlight today, but Hiroko has a day just about the same as ours. It's pretty hard to figure out."

At that moment, the disk of the Sun touched the horizon. Wendy and her dad watched in silence until there was just one brilliant sliver remaining. "Going...going...gone," she whispered. "It must be 8:37 p.m."

As they started back down to their apartment, Wendy asked, "What causes days to get longer in some places, shorter in other places, and stay the same in still other places?"

"The answer to that question is also the key to understanding the seasons. Look for the answer not in what the Sun is doing, but what Earth is doing. You can shed a little light on the subject by looking closely at the revolution of Earth around the Sun and the rotation of Earth on its axis."

"Very funny, Dad, but thanks for the tip."

Think Questions

1. Which locations have the greatest number of hours of daylight on June 21? The fewest hours of daylight?
2. Which locations have the longest hours of daylight on December 21? The shortest hours of daylight?
3. Alpena, Michigan, is located 45° north of the equator. How much daylight do you estimate they have on June 21? On December 21?
4. Boulder, Colorado, has a latitude of 40°N. Wellington, New Zealand, has a latitude of 41°S. Which city has the longest amount of daylight on June 21?

Seasons

What do you picture in your mind when you read these words? Summer. Fall. Winter. Spring.

Most of us come up with a mental picture or two—summer means shorts and T-shirts, swimming, and fresh fruits and vegetables. Winter means heavy coats and short days with, perhaps, a blanket of snow on everything. Seasons are pretty easy to tell apart in most parts of the country. The amount of daylight, the average temperature, and the behavior of plants and animals are a few familiar indicators of the season. But what causes the predictable change of season? Have you ever stopped to think about why the seasons happen?

As Earth Tilts

Let's start with a quick review of some basic information about our planet.

- Earth spins on an imaginary axle called an **axis.** The axis passes through the North and South Poles. This spinning is called **rotation.** It takes 24 hours for Earth to make one rotation on its axis.
- Earth travels around the Sun. Traveling around something is called **revolution.** Earth's path around the Sun is not exactly round, but is slightly oval. One revolution takes 365 and 1/4 days, which is 1 year.

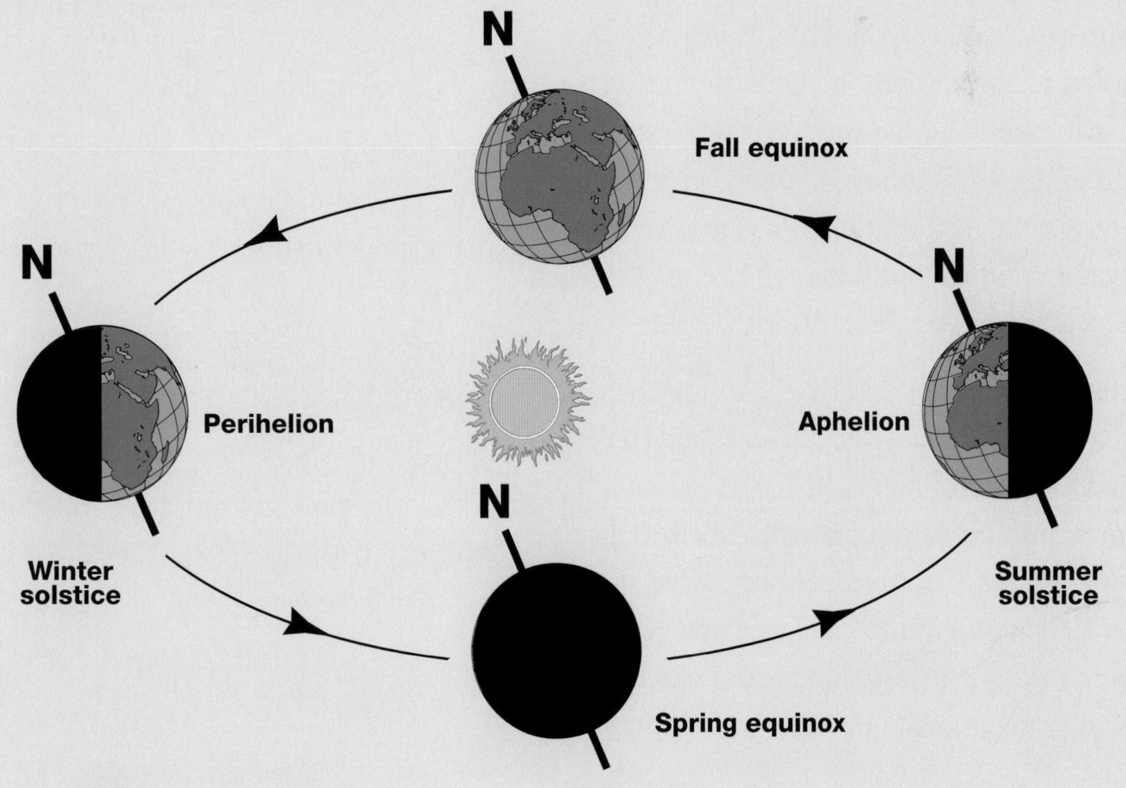

- Earth doesn't sit straight up and down on its axis as it revolves around the Sun. It is tipped at a 23.5° angle.
- The average distance between the Sun and Earth is about 150 million kilometers. Because Earth's orbit is an ellipse (oval), Earth is sometimes farther away from and sometimes closer to the Sun. **Perihelion** is when Earth and the Sun are closest to each other. Perihelion happens each year around January 3. The distance is 147 million kilometers. **Aphelion** is when Earth and the Sun are farthest apart. It happens each year around July 4. The distance is 152 million kilometers.

It would seem logical that summer would be during perihelion, when Earth is closest to the Sun. Wrong. Here in the Northern Hemisphere, we are in the middle of winter at the time of perihelion. Because Earth is closest to the Sun in January, it receives more energy in January than at any other time of year. But that energy doesn't make it warm in the United States. The reason for seasons is linked to Earth's tilt.

Think about Earth revolving around the Sun. As Earth revolves, it also rotates on its axis, one rotation every 24 hours. Here's something important: Earth's North Pole *always* points at a reference star called the North Star. No matter where Earth is in its orbit around the Sun, the North Pole always points at the North Star, day and night.

Tilt Equals Season

Look at the illustration on page 17. It shows where Earth is in its orbit around the Sun at each season. You will also see that the North Pole points toward the North Star in all four seasons.

Study the Earth image in the summer solstice position. Because of the tilt, Earth is "leaning" toward the Sun. When the North Pole is leaning toward the Sun, it is summer in the Northern Hemisphere. Days are longer, and the angle at which light hits that part of Earth is more direct. Both of these factors result in more solar energy falling on the Northern Hemisphere in summer (thus more heat) even though the planet is actually farther away from the Sun.

Look at the position of Earth 6 months later (winter solstice). Just the opposite is true. Even though Earth is closer to the Sun at this time, the Northern Hemisphere is leaning *away from* the Sun. Days are shorter, and sunlight does not come as directly to the Northern Hemisphere, so it gets less solar energy.

Four days in the year have names based on Earth's location around the Sun. **Summer solstice** (June 21 or 22) is the day when the North Pole leans toward the Sun. **Winter solstice** happens on December 21 or 22 when the North Pole leans away from the Sun.

The 2 days when the Sun's rays shine straight down on the equator are the **equinoxes.** Earth's axis is tilted neither away from nor toward the Sun. *Equinox* means "equal night." Daylight and darkness are equal (or nearly equal) all over Earth. There are two equinoxes each year, **spring equinox** (March) and **fall equinox** (September).

Daily Dose of Sunshine

We take night and day for granted. They always happen. The Sun comes up; the Sun goes down. This cycle has happened as long as humans have been on Earth. It will most likely continue for millions of years.

Because Earth tilts, the length of day and night changes as the year passes. This table shows how hours of daylight change by latitude during the year. When it's summer in the Northern Hemisphere, the North Pole leans toward the Sun. At the North Pole, the Sun never sets. Above the Arctic Circle (66.5° north), daylight can last all 24 hours of the day.

LENGTH OF DAYLIGHT IN THE NORTHERN HEMISPHERE			
Latitude (°N)	Summer solstice	Winter solstice	Equinoxes
0	12 hr.	12 hr.	12 hr.
10	12 hr. 35 min.	11 hr. 25 min.	12 hr.
20	13 hr. 12 min.	10 hr. 48 min.	12 hr.
30	13 hr. 56 min.	10 hr. 04 min.	12 hr.
40	14 hr. 52 min.	9 hr. 08 min.	12 hr.
50	16 hr. 18 min.	7 hr. 42 min.	12 hr.
60	18 hr. 27 min.	5 hr. 33 min.	12 hr.
70	24 hr. 00 min.	0 hr. 00 min.	12 hr.
80	24 hr. 00 min.	0 hr. 00 min.	12 hr.
90	24 hr. 00 min.	0 hr. 00 min.	12 hr.

THERMOMETER: A Device to Measure Temperature

Is the oven ready for this pie? You look flushed—do you have a fever? The fish are not eating. Is it warm enough in the aquarium? The ice cream is soft. Is that freezer working?

To answer these questions, we reach for a thermometer. And these days, there are lots of different kinds to reach for.

All thermometers work the same way on a basic level: some property of a material changes as it gets hot. You may already be familiar with several kinds of thermometers. The old standby is the glass tube filled with alcohol or mercury. This is how it works.

A thin, heat-tolerant, glass tube with a bulb at one end is filled with alcohol or mercury. The liquid extends partway up the tube. The tube is then sealed and attached to a backing that has a scale written on it.

When the bulb touches something hot, the liquid inside expands. The volume of liquid increases. The only place the added volume of liquid can go is into the tube. The distance the liquid pushes up the tube indicates the temperature of the material touching the thermometer bulb.

The first closed-tube thermometer, like the one described above, was invented by Grand Duke Ferdinand II in 1641. He used alcohol in the tube. During the 18th century, more precise closed-tube thermometers made it possible to conduct experiments involving fairly accurate temperature measurements.

In 1714, German physicist Daniel Gabriel Fahrenheit made a mercury thermometer and developed the Fahrenheit temperature scale. On the Fahrenheit scale, 32°F is the freezing point of water and 212°F is the boiling point of water. In 1742, Swedish astronomer and physicist Anders Celsius devised a temperature scale on which 0°C is the freezing point of water and 100°C is the boiling point. This used to be called the centigrade scale, which means "hundred steps." But in 1948, it was renamed the Celsius scale in honor of Anders Celsius.

There are other types of thermometers. Oven thermometers and some wall thermometers look a little like pocket watches. Inside is a **bimetallic strip.** Bimetallic strips are made of two metals stuck together. The two metals expand at different rates when they get hot. When the heat is turned up, the copper-colored part of the strip expands (lengthens) more than the other. The strip bends. A pointer attached to the bending metal strip points to the temperature.

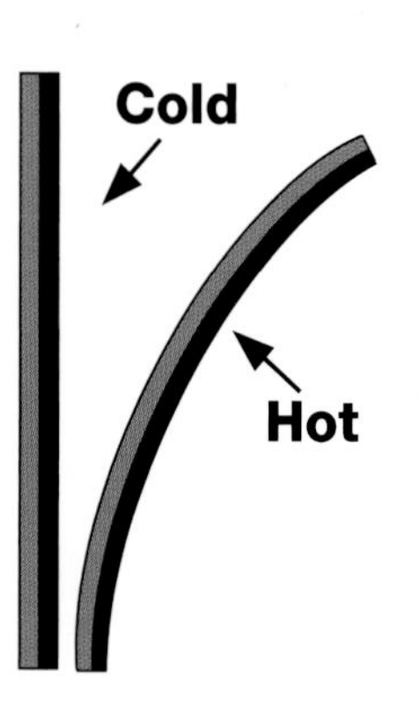

Tropical-fish fanciers keep thermometers right in the aquarium. One efficient kind is a thin, flat strip of plastic, like a piece of black plastic ribbon, that has **liquid crystals** packaged inside. Liquid crystals change color within a very narrow temperature range. A liquid-crystal thermometer has a series of little pockets in the strip, each filled with a different mix of liquid crystals to indicate one temperature only. So all you have to do is look at the strip to see which number is surrounded by a green glow, and that's the temperature.

22 23 24 25 26 27 28

30 31 32 33

The last time you went to the doctor for a checkup, you may have had your temperature taken with a digital thermometer. These recent arrivals on the thermometer scene are very accurate and easy to use. You slip the probe end under your tongue for a few seconds. Inside the probe is a circuit with electricity flowing through it. Part of the circuit flows through a piece of wire that changes resistance as the temperature increases. When the electronic circuitry detects that the current flowing in the probe circuit has stabilized, that means the temperature is no longer changing. The electronic thermometer measures the amount of current flowing in the circuit and displays the temperature on a little digital screen.

That's just a small sampling of the many different thermometers found in common and specialized applications.

Galileo *invented one of the first functional thermometers in 1596. He filled a number of small glass balls partway with colored water and sealed them shut. The balls of colored water all floated in water.*

Galileo knew that water expands as it warms up. Warm, expanded water is less dense than cold water. He then attached a weight to each ball. The weights were adjusted to give each ball a slightly different buoyancy. The result was that when the water was cold (at its densest), all of the balls floated. As the water warmed up, becoming less dense, balls would sink.

By placing the balls in a column of water in order of their buoyancy, with the least dense on the top, Galileo produced a thermometer. If all of the balls were on the bottom of the cylinder, it was really hot.

Modern versions of the Galileo thermometer have temperatures printed on the weights. The number on the lowest floating ball shows the temperature of the system.

Heating the Atmosphere

You may have had an experience like this one. The campfire has burned down to a bed of hot coals, perfect for toasting some marshmallows. The only stick available is about a meter long, but you go for it. You can hardly stand the heat from the coals because the stick is short, but after a minute the marshmallow is brown and gooey. You pop it into your mouth. Yikes! Didn't wait long enough for it to cool.

This story includes a couple of intense heat experiences. But have you ever stopped to think about what heat really is? What is the heat that you felt coming off the coals and the heat in the marshmallow that burned your tongue?

Heat = Movement

Objects in motion have energy. The faster they move, the more energy they have. Energy of motion is called **kinetic energy.**

Matter, like nails, soda bottles, water, and air, is made of atoms and molecules. Atoms and molecules, even in steel nails and glass bottles, are in motion. In solids, the molecules vibrate back and forth. In liquids and gases, the molecules move all over the place. The faster molecules vibrate or move, the more energy they have.

Molecular motion is molecular kinetic energy, and that is heat. The amount of kinetic energy in the molecules of a material determines how much heat it has. The molecules in hot materials are moving fast. The molecules in cold material are moving slowly.

Heat Transfer

Heat can move, or transfer, from one place to another. Scientists sometimes describe heat transfer as heat flow, as though it were a liquid. Heat is not a liquid, but flow is a pretty good way to imagine its movement.

Heat flows from a hotter location (more energy) to a cooler one (less energy). For example, if you add cold milk to your hot chocolate, heat flows from the hot chocolate to the cold milk. The hot chocolate cools because heat flows away; the cold milk warms because heat flows in. Soon the chocolate and the milk arrive at the same temperature, and you gulp them down.

Heat Transfer by Radiation

There are many different forms of energy, including heat and light. If you heat an object, like the burner on a stove, to a high enough temperature, it will get red hot. When this happens, the burner is giving off two forms of energy, heat and light. If you put your hands near a lightbulb, you can see light and feel heat, even though you are not touching the bulb. This kind of energy that travels right through air is **radiant energy.**

Radiant energy travels in the form of **rays.** Heat and light rays radiate from sources like the intensely hot campfire coals, lightbulbs, and the Sun.

Energy rays from the Sun pass through Earth's atmosphere. We call this **solar energy.** When solar energy hits a molecule, such as a gas molecule in the air, a water molecule in the ocean, or a molecule in the soil, the energy can be **absorbed.** Absorbed energy increases the kinetic energy (movement) of the molecules in the air, water, or soil. Increased kinetic energy equals increased heat.

Radiation is one way energy moves from one place to another. Materials don't have to touch for energy to transfer from one to the other.

Heat Transfer by Conduction

Think about that hot toasted marshmallow or maybe a slice of pizza straight from the oven going into your mouth. This kind of memorable experience is another kind of energy transfer. When energy transfers from one place to another *by contact,* it is called **conduction.**

The fast-moving molecules of the hot pizza bang into the slower molecules of your mouth. The molecules in your tongue gain kinetic energy. At the same time, molecules of the hot pizza lose kinetic energy, so the pizza cools off. Some of the pizza kinetic energy is conducted to heat receptors on your tongue, causing them to send a message to your brain that says "Hot!"

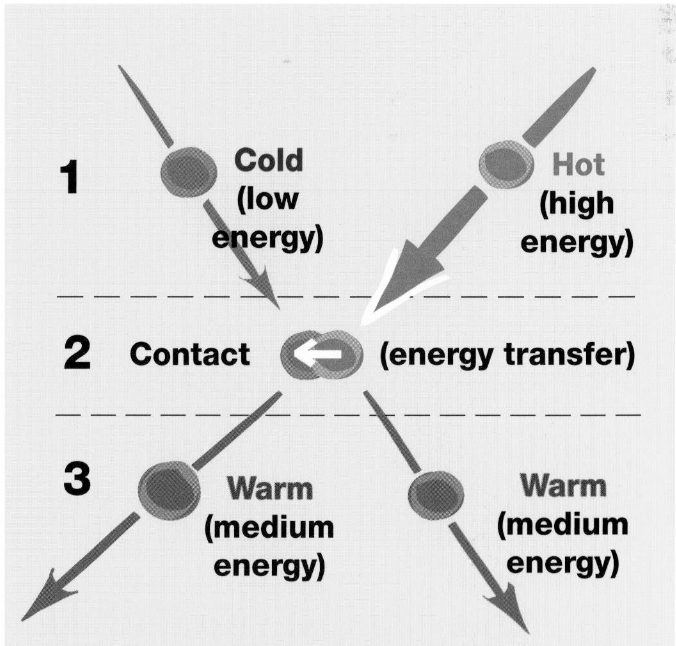

A "hot" molecule with a lot of kinetic energy collides with a "cold" molecule with little kinetic energy. Energy transfers at the point of contact. The cold molecule then has more energy, and the hot molecule has less energy.

When you heat water in a pot, the water gets hot because it comes in contact with the hot metal of the pot. Kinetic energy transfers from the hot metal molecules to the cold water molecules by contact, which is conduction.

Heat Transfer to the Atmosphere

The atmosphere is heated by radiant energy from the Sun—solar energy. Lots of different kinds of rays are sent out by the Sun, but the most important ones are visible light and invisible light called infrared radiation. It seems pretty straightforward. The molecules in the air absorb the incoming radiation to increase their kinetic energy. But that's not what happens.

Air is 99% nitrogen and oxygen molecules. Neither of these molecules absorbs visible light or infrared radiation. It just doesn't happen. Only water vapor and carbon dioxide absorb significant amounts of radiant energy, and this is mostly infrared rays, not visible light.

If only a tiny part of the atmosphere gets hot from incoming solar energy, how does the rest of the atmosphere get hot?

Visible light *is* absorbed by Earth's surface. The land and seas warm up. The air molecules that come in contact with the warm land and water molecules gain energy by conduction. But there is more.

The warm land and seas also **reradiate** energy. This is a very important idea. Earth actually gives off infrared radiation that can be absorbed by water molecules (mostly) and carbon dioxide molecules in the atmosphere. The energy absorbed by the small number of water molecules is transferred throughout the atmosphere by conduction when hot water molecules bang into oxygen and nitrogen molecules.

The atmosphere is not heated from above; it is heated from below.

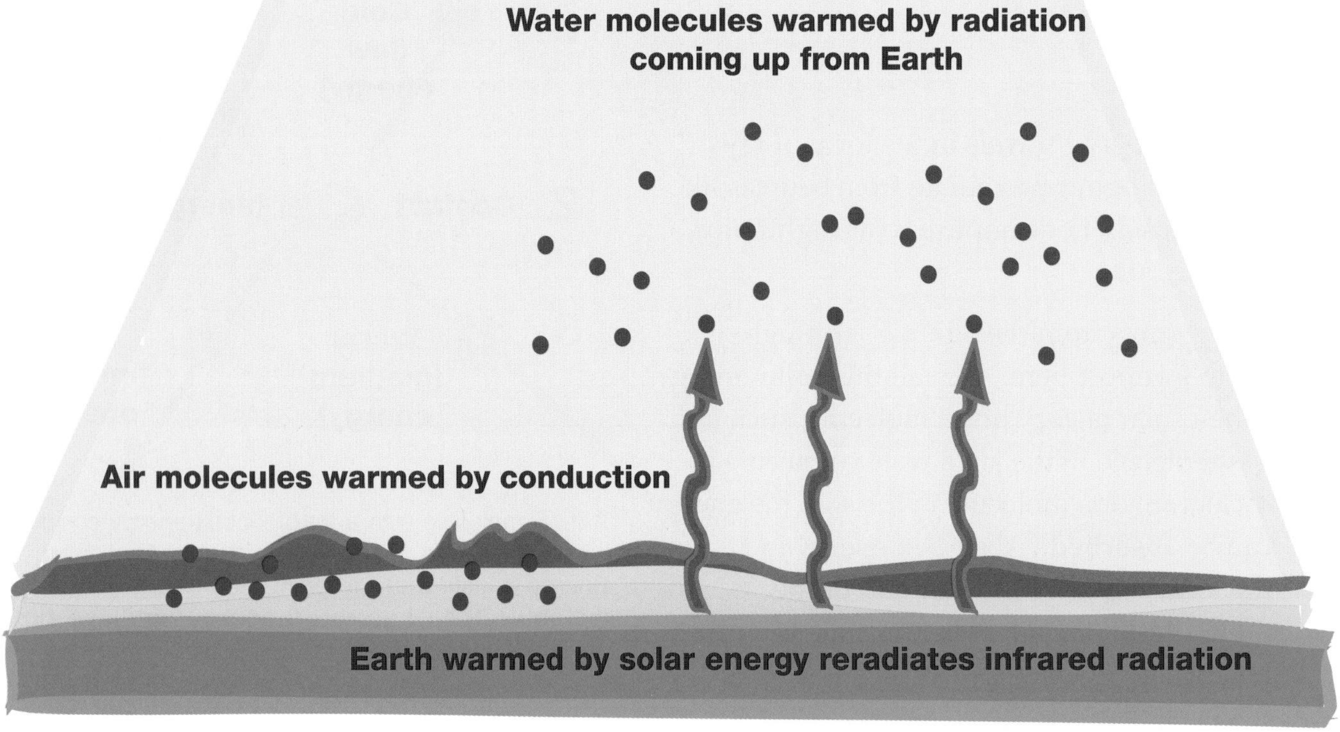

Temperature and Thermometers

How can you find out just how much heat is in the part of the atmosphere where you are? With a thermometer.

A thermometer measures temperature. **Temperature** is a measure of the average kinetic energy of the molecules in a material. If a thermometer is surrounded by air, it measures the average kinetic energy of the air molecules. If it is surrounded by water, it measures the kinetic energy of the water molecules. If you hold the thermometer bulb between your fingers, the thermometer measures the average kinetic energy of the molecules on the surface of your fingers.

If you stick a thermometer in a cup of cocoa, under your tongue, or in a freezer, it will measure the average kinetic energy of the molecules touching it in those places.

How Does a Thermometer Work?

Think about an alcohol thermometer on the wall in a cold cabin. The kinetic energy in the air molecules is low. The kinetic energy in the glass and alcohol molecules is low. The air molecules and the glass thermometer bulb have the *same* kinetic energy. The top of the column of alcohol is at 5°C. Brrrr, it's cold, so you turn on the heater.

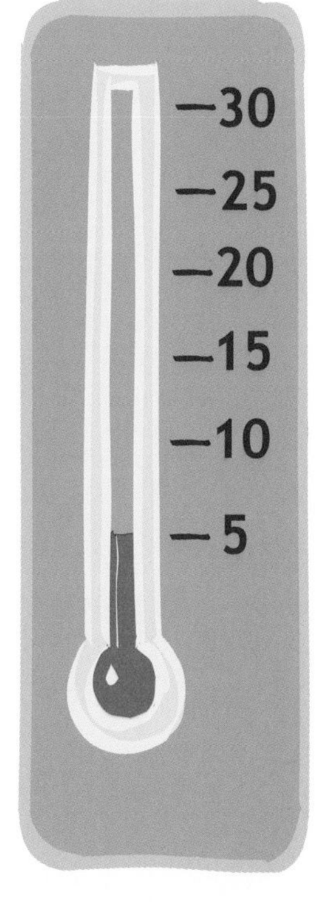

Pretty soon warm air is flowing into the room. Warm air has more kinetic energy than cooler air. The energy added to the room in the form of fast-moving air molecules starts a chain of events.

- Molecules in the warm air collide more often with the glass thermometer bulb. Energy transfers to the molecules of glass by conduction.

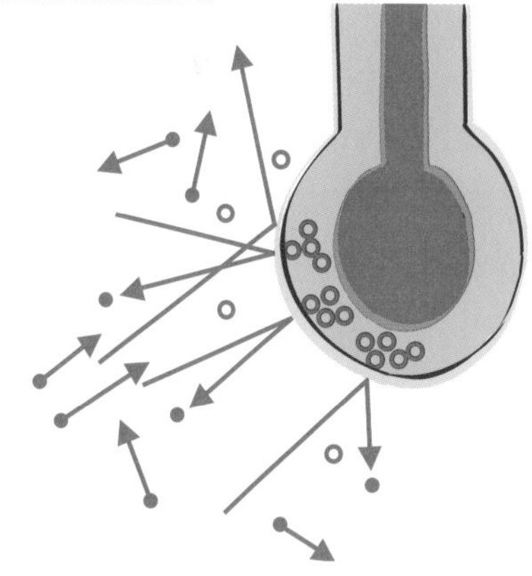

- Kinetic energy transfers by conduction from molecule to molecule in the glass bulb until all of the glass molecules are warm.

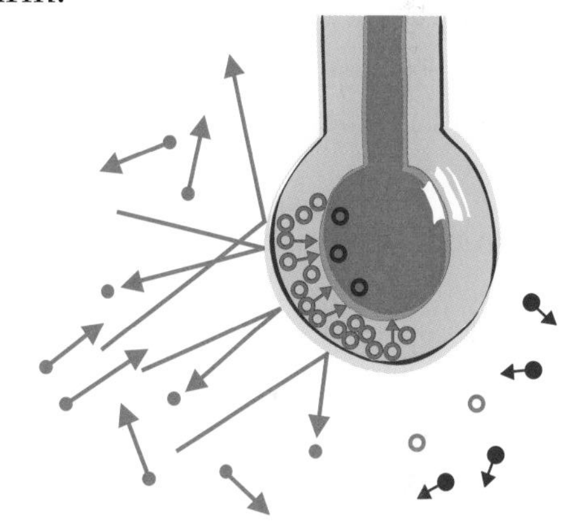

- Energy transfers by conduction from the glass molecules to the alcohol molecules inside the bulb. Kinetic energy transfers throughout the alcohol by conduction—collisions between alcohol molecules.

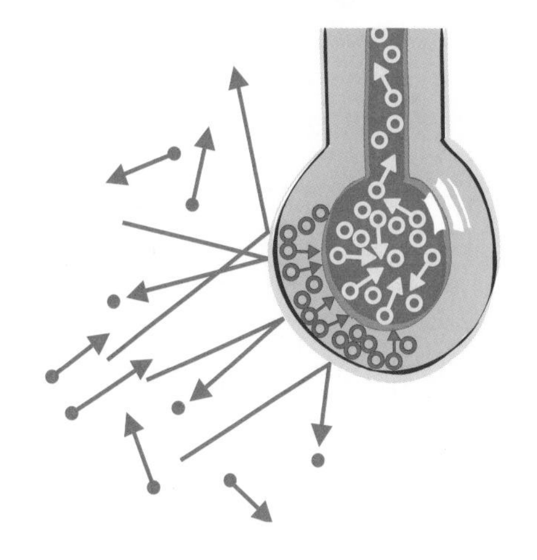

- The alcohol molecules push on one another more forcefully because of the increased kinetic energy. The molecules push farther apart, expanding the volume of alcohol. As the volume of alcohol gets bigger, the only place it can go is into the tube.
- The alcohol will continue to expand as long as more and more energy is transferred to the alcohol. The alcohol will continue to push up the tube.

When the room is warm, you turn off the heater. In a few minutes, all the molecules in the room are at the same level of kinetic energy. The alcohol stops rising. The top of the column of alcohol is right next to the 20°C mark on the thermometer. Nice and warm.

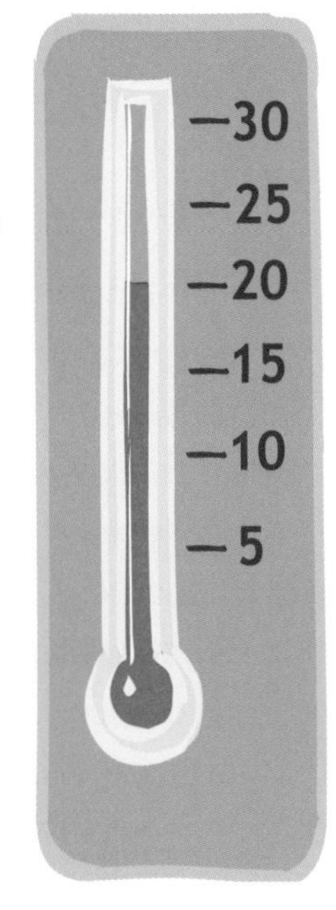

As long as the kinetic energy of the alcohol molecules stays the same, the level of the column of alcohol will stay the same, and we say the temperature is steady at 20°C.

What happens if you open the window? The whole process goes into reverse. Molecules in cool air have less kinetic energy. Heat energy transfers by conduction from the glass tube to the air. Heat then transfers from the alcohol to the now-cooler glass. The alcohol molecules lose kinetic energy and slow down, and the alcohol contracts. The liquid level gets lower in the tube. Kinetic energy will transfer from the molecules in the room to the molecules outside through the open window. The chill of low kinetic energy will set in once again. The thermometer will once again dip to 5°C. Brrrr.

Think Questions

1. What is heat?
2. What heats Earth's atmosphere?
3. What are the two ways energy can transfer from one material to another?
4. Explain two ways that Earth's atmosphere gets heated.
5. Thermometers measure temperature. What is temperature?
6. Explain why the alcohol level in a thermometer goes down when the weather gets cold.

DENSITY

Make believe you have a package of regular rubber balloons. Fill one with water, tie it off, and give it to a friend. Fill a second, identical balloon with air until it is the same size as the water balloon. Tie it off and give it to a second friend to hold. Fill a third balloon with helium, same size as the other two, and tie it off.

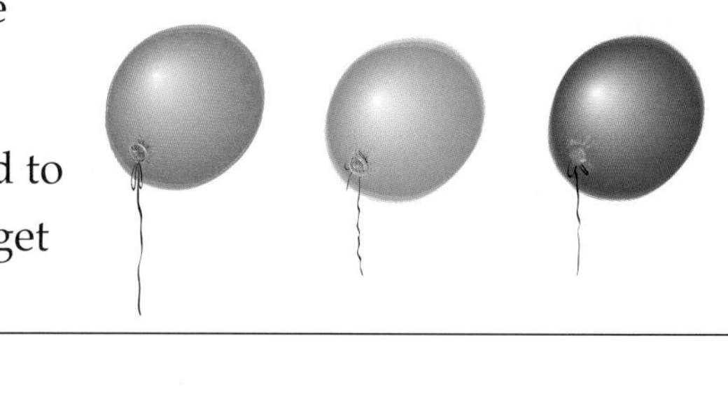

Review the balloons—three identical balloons, all filled to exactly the same volume, each tied off so nothing can get in or out. What's different? The kind of material in the three balloons. Ready to try a little experiment?

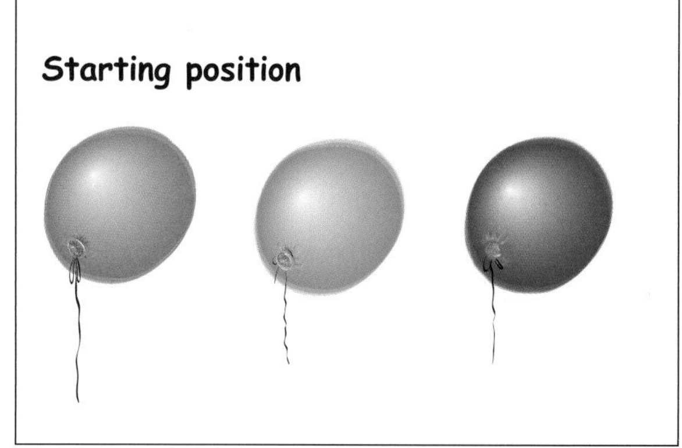

You and your friends hold the balloons at the same height above the floor. On the count of three, you will all release your balloons and observe what happens.

The water balloon will plunge to the floor, the air balloon will drift slowly to the floor, and the helium balloon will float up to the ceiling. Why?

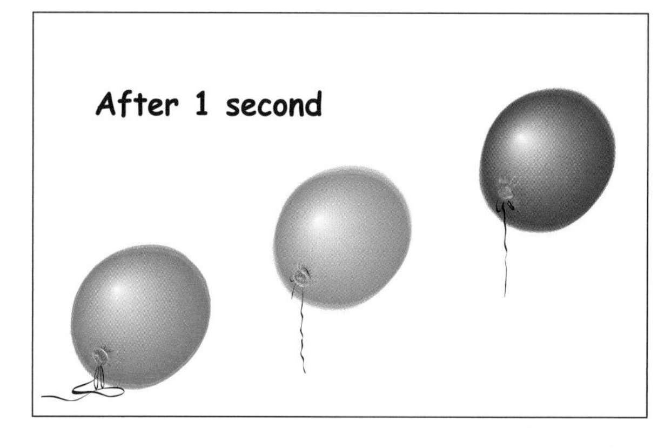

It comes down to how much stuff there is in each balloon. The scientific word for stuff is **matter.** The amount of matter in an object is its **mass.** Matter is made out of atoms. So the mass of an object depends on *how many* atoms there are in the object and *how big* the atoms are.

The atoms in solids (glass, steel) and liquids (water, alcohol) are packed together as close as they can get. This means there are lots of atoms in a volume of water. That makes water pretty heavy.

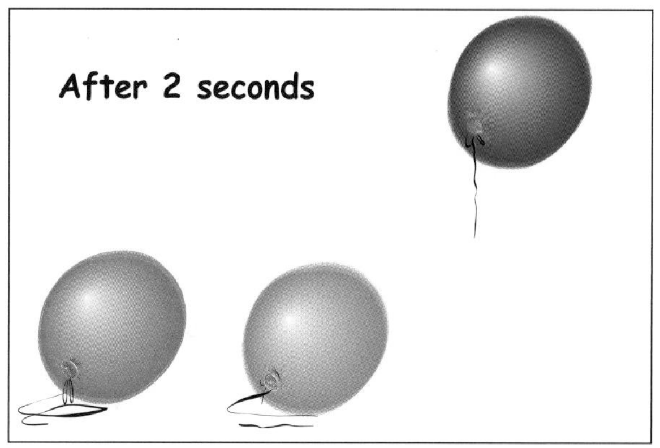

In gases, the atoms are not packed as close together as they can get. There is a lot of space between atoms. Air and helium are gases, so they are pretty light.

Air atoms (mostly nitrogen and oxygen) are fairly large, but helium atoms are small. Generally speaking, small atoms weigh less than large atoms. So helium is much lighter than air.

DENSITY

The amount of matter in a volume of material determines the material's density. **Density** is defined as mass per volume of an object. When you have equal volumes of a bunch of different materials, you can find out which one is densest and which one is least dense by weighing them. The heaviest one is the densest; the lightest is the least dense.

The important idea in this discussion is that you need to compare the weight of *equal volumes* of different materials to determine which one is densest.

DENSITY OF LIQUIDS

Mr. Dey's students had several salt solutions. They wanted to find out which one was densest.

Group 1 put 25 milliliters (ml) of the blue solution on a scale and found that it had a mass of 25 grams (g). They measured 25 ml of green solution into another cup. Its mass was 30 g.

The students announced, "We weighed equal volumes of two solutions. The green solution is heavier, so it is denser. It has more mass per volume than the blue solution."

Group 2 put 25 ml of blue solution on a scale and found that it had a mass of 25 g. But then they made a little mistake. They put 50 ml of yellow solution in a cup and found that it had a mass of 55 g.

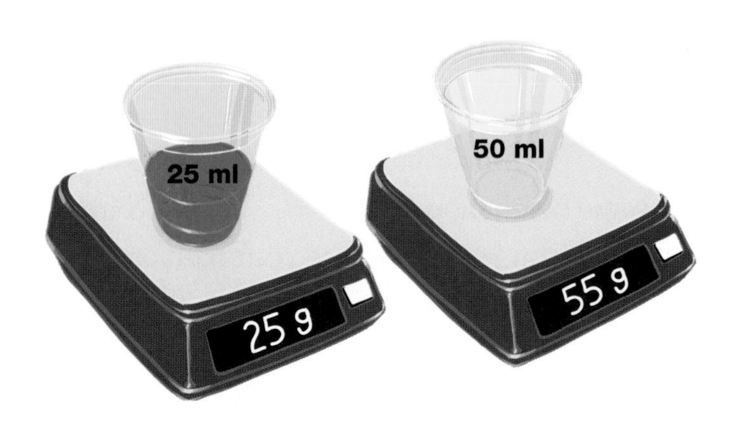

When they realized what they had done, Reggie said, "Oh-oh, we didn't measure equal volumes. We have to start over."

"Maybe not," said Yolanda. "We weighed twice as much yellow solution as we should have. If we had used half as much, it would have weighed half as much. All we have to do is divide the mass by two to find out the mass of 25 ml of yellow solution."

They did the math and found that 25 ml of yellow solution had a mass of 27.5 g.

The two groups put their data together in a table.

Solution	Volume	Mass
Blue	25 ml	25 g
Green	25 ml	30 g
Yellow	25 ml	27.5 g

Students could now easily compare equal volumes of the three solutions to see which one was heaviest and, therefore, densest. They determined that green was densest, blue least dense, and yellow in the middle.

Mr. Dey had a question. "What is the mass of 1 ml of each of the solutions?"

Reggie offered, "Twenty-five milliliters of blue solution has a mass of 25 g, so 1 ml of blue has a mass of 1 g."

"And the green solution?" asked Mr. Dey.

Reggie's group thought about it this way.

- Twenty-five milliliters of green has a mass of 30 g. That's more than 1 g for each milliliter.
- They drew a pie chart to help them think about the problem. Each slice of pie represented 1 ml, or 1/25 of the total volume.

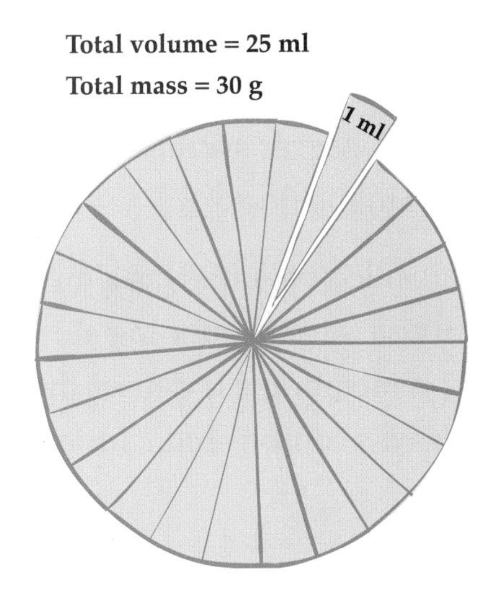

- One milliliter is 1/25 of the total volume. Each milliliter must have 1/25 of the total mass.
- They divided 30 g by 25 ml to find the mass of 1 ml of green solution.

The students discovered the definition of density. **Density is the mass, in grams, of 1 ml of material.**

The usual way of stating density is *mass per volume.* The word *per* means "divided by."

Density can be written as an equation.

$$\text{Density} = \frac{\text{Mass}}{\text{Volume}} = \frac{\text{g}}{\text{ml}}$$

The equation can be used to calculate the density of the green solution.

$$\text{Density} = \frac{\text{Mass}}{\text{Volume}} = \frac{30\text{ g}}{25\text{ ml}} = 1.2\text{ g/ml}$$

Density equals mass divided by volume. If you remember "mass per volume," you will always know how to set up your equation when it comes time to calculate a density.

What's the density of the yellow solution? Remember, mass per volume. The original mass of the yellow solution was 55 g, and the volume was 50 ml.

$$\text{Density} = \frac{\text{Mass}}{\text{Volume}} = \frac{55\text{ g}}{50\text{ ml}} = 1.1\text{ g/ml}$$

DENSITY INTERACTIONS

Density is a number that tells you how much matter there is in a milliliter or **cubic centimeter (cc)** of material. One milliliter is exactly the same volume as one cubic centimeter.

Liquids are generally measured in milliliters, and solids and gases are measured in cubic centimeters.

If there is a lot of matter in a cubic centimeter of something, it is dense. If there is very little matter in a cubic centimeter of material, it is not dense.

Water has a density of 1 g/ml. Materials with densities greater than 1 g/ml are denser than water; materials with densities less than 1 g/ml are less dense than water.

What will happen if you put a rock with a density of 3 g/cc in a tub of water? It will sink like, well, a rock. And if you put a cork, with a density of 0.45 g/cc in the tub of water? Yes, it will float.

Materials that are denser than water sink in water. Materials that are less dense than water float in water. That's the way it always works.

When salt dissolves in water, it forms a solution. The more salt dissolved in a volume of water, the greater the solution's density.

If you carefully pour a little bit of each of the colored solutions from Mr. Dey's class into a vial, what do you think would happen? Can you describe the result?

Solution	Density
Blue	1.0 g/ml
Yellow	1.1 g/ml
Green	1.2 g/ml

DENSITY OF GASES

Back to the balloons. We know the density of water, but what about the air and the helium?

Material	Density
Water	1.0 g/ml
Air	0.0013 g/ml
Helium	0.0002 g/ml

The chart shows that air and helium are not very dense. There is very little mass in a milliliter of gas.

When the three balloons were released, the dense water-filled balloon pushed through the less-dense air surrounding it. It fell straight to the floor.

The air-filled balloon was almost the same density as the surrounding air. The rubber-balloon membrane is denser than air, and the air was compressed a little inside the balloon, making the air-filled balloon a little denser than the surrounding air. It drifted slowly to the floor.

The helium-filled balloon was quite a bit less dense than the surrounding air. Just like a cork in water, the less-dense helium balloon floated up to the ceiling.

Density of Air

Air is gas. The molecules in gases are not bonded to other molecules. Gas molecules move around freely in space.

When energy transfers to matter, the kinetic energy (movement) of the atoms and molecules increases. The increased motion causes most matter to expand. When matter expands, the atoms and molecules do not get bigger—they get farther apart. This is a very important point: It is the distance between molecules that increases, not the size of the molecules.

When matter expands, the molecules get farther apart. What do you think that does to the density of the material? The density gets lower. When the molecules get farther apart, each cubic centimeter (which is the same as a milliliter) has fewer molecules. Fewer molecules per milliliter means lower density.

This is a general rule of matter. When matter gets hot, it expands, and the density goes down.

Air is matter. Air expands when it gets hot. Air gets less dense when it gets hot. When energy transfers from Earth's surface to the air by conduction (contact between surface molecules and air molecules) or reradiation, the air temperature goes up and the air expands. The low-density, warm air rises, just like the helium balloon in air.

Density and Weather

Weather happens in the atmosphere. Energy transfers into and out of the atmosphere in the form of heat. As air heats up and cools down, its density changes. Warm air tends to go up, and cold air tends to go down. When masses of air move, things happen in the weather.

The idea of density will be an important concept in our investigation of weather and the processes that cause it.

Think Questions

1. **What do you think the density of a person might be? Explain.**
2. **Why do you think hot-air balloons are able to rise into the air? How do hot-air-balloon pilots get their balloons back to Earth?**

CONVECTION

If you put a couple of centimeters of water in a metal pie pan and support it over a candle, you can slowly heat the water. The energy from the flame will heat a small area of the pan, and the heat will conduct to the water in contact with the hot metal.

A small mass of water will heat up. The question is, what happens to the hot water?

The water expands as the water molecules gain kinetic energy and push farther apart. The expanding water is less dense than the surrounding water. The warm water rises upward.

When the warm water reaches the surface, it spreads out. Water at the surface cools by radiation and conduction.

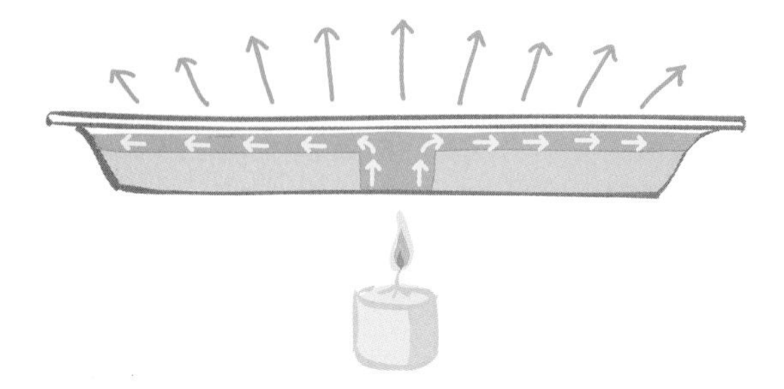

When the cool water reaches the edges of the pie pan, the water is dense. It flows down the sides of the pan, across the bottom, and back toward the center of the pan. As the water nears the hot metal, it begins to warm again. The hot water rises to repeat the cycle.

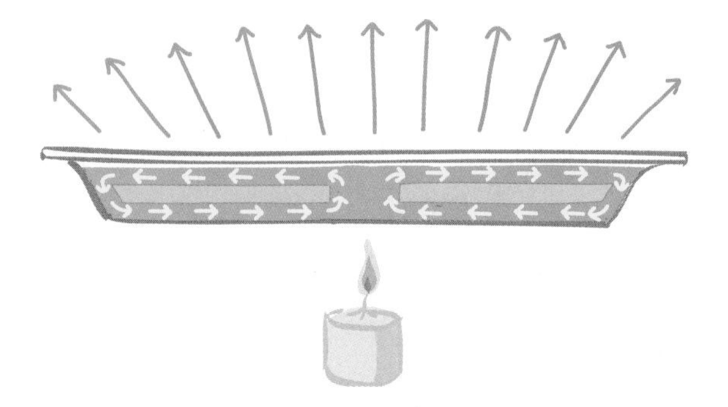

The movement of water in the pan, driven by a localized heat source, is **convection.**

Convection happens only in fluids. Fluid near an energy source heats up and expands. The hot fluid becomes less dense and rises. The energy in the molecules of hot fluid is carried to a new location. As the energy in the hot fluid transfers away from the fluid, it cools and contracts, making it denser. The cool fluid flows downward again.

The mass of fluid flowing in a circle is called a **convection cell.**

Air is a fluid. Energy can transfer to air, causing it to expand. When air expands, it becomes less dense. Less-dense air will rise in the atmosphere.

This is exactly what happens in the real world. Earth's surface is always warm in the tropics, the part of the planet near the equator. Water in the tropical oceans absorbs a lot of solar energy. Air in contact with the tropical seas receives a lot of energy by reradiation and conduction. Huge masses of air heat up and begin to rise. This is the start of the largest convection cell on Earth.

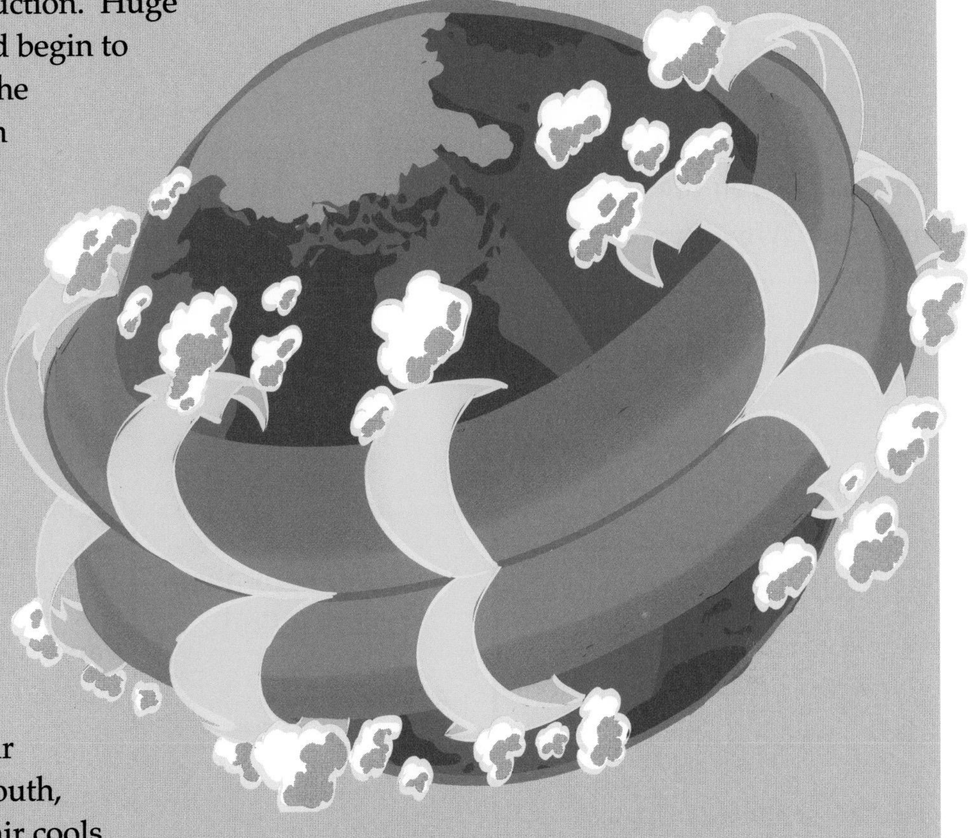

The equatorial convection cells circle the globe like two bicycle inner tubes. Because much of the energy transfer occurs over the ocean, large amounts of water vapor rise high into the atmosphere, riding along in the convection cell. The warm, moist air spreads out north and south, and it cools. When the air cools, water vapor condenses into droplets of liquid water. Large numbers of little droplets of water form clouds. And we all know what happens after clouds form—rain.

In the next few investigations, we will see how the process of convection helps redistribute water around the planet and affects wind, everything from gentle breezes to powerful, dangerous storms.

Dragon's Breath

Did you ever breathe out a big breath of steam, just like a dragon? It's great fun to exhale a cloud in front of your face and pretend that you can scorch the countryside with one blast of your mighty breath. But you can't perform this trick all the time. Why can you see these breath clouds sometimes, but not all the time?

That short-lived dragon's breath is a little cloud. What are the ingredients you need to make a cloud? You need water vapor in the air, temperature at or below dew point, and a surface on which the water vapor can condense. Let's look at the variables one at a time.

Water Vapor in the Air

There is always water vapor in the air. Sometimes there is very little water vapor, and sometimes there is a lot. Water vapor in the air is called **humidity.**

Temperature plays an important role in humidity. The rule is, the warmer the air, the more water vapor it can hold. At 35°C, a kilogram (kg) of air can hold 35 grams (g) of water vapor. That same kilogram of air at 0°C can hold only 3.5 g of water vapor.

When air is holding as much water vapor as it can, it is said to be **saturated.** When air is saturated, no more water vapor can enter the air.

The amount of water needed to saturate a mass of air is not the same at all temperatures. Let's think about a kilogram of air that is holding 3.5 g of water vapor. The 3.5 g of water vapor will saturate the kilogram of air at 0°C. That same 3.5 g of water vapor, however, represents only 10% of the 35 g needed to saturate the kilogram of air at 35°C. The amount of water vapor in the air compared to the amount of water vapor *needed to saturate the air at a given temperature* is **relative humidity.** Relative humidity is reported as a percentage.

The relative humidity of the kilogram of air holding 3.5 g of water vapor at 0°C is 100%. When the same kilogram of air with the same 3.5 g of water vapor is heated to 35°C, the relative humidity is only 10%.

Water/air saturation points (g/kg)

Air temp. (°C)	Grams water per kilogram of air
–40	0.1
–30	0.3
–20	0.8
–10	2.0
0	3.5
5	5.0
10	7.0
15	10.0
20	14.0
25	20.0
30	26.5
35	35.0
40	47.0

Dew Point

Picture the kilogram of warm 35°C air holding 3.5 g of water vapor. That's 10% relative humidity. This is a cartoon of 3.5 g of water vapor in the air compared to the 35 g of water vapor needed to saturate the air at 35°C.

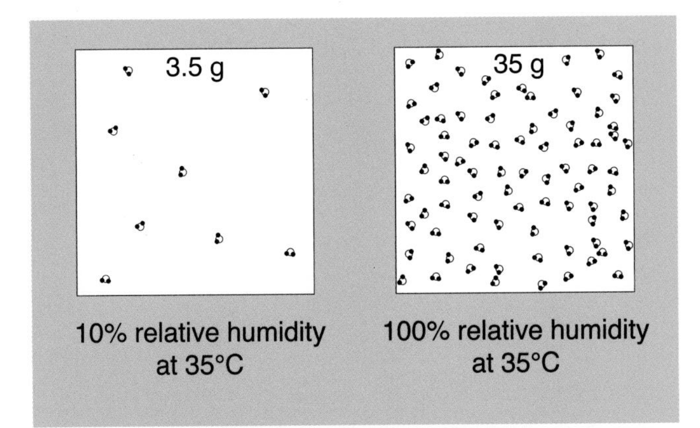

10% relative humidity at 35°C

100% relative humidity at 35°C

As the kilogram of air cools off to 20°C, there are still 3.5 g of water vapor in the kilogram of air, but at 20°C a kilogram of air is saturated when it is holding 14 g of water vapor.

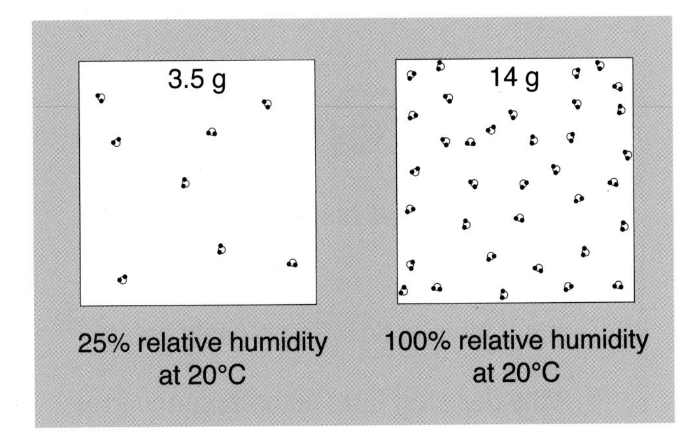

25% relative humidity at 20°C

100% relative humidity at 20°C

At 10°C, a kilogram of air is saturated when it is holding 7 g of water vapor. When our kilogram of air with its 3.5 g of water vapor cools to 10°C, the relative humidity is now up to 50%.

If the mass of water vapor in a kilogram of air doesn't change, its relative humidity goes up and up as it cools. When it gets to 0°C, an interesting thing happens. Without changing the mass of the water vapor in the air, the air becomes saturated. The kilogram of air can hold only 3.5 g of water vapor at 0°C, and that's the amount in our air.

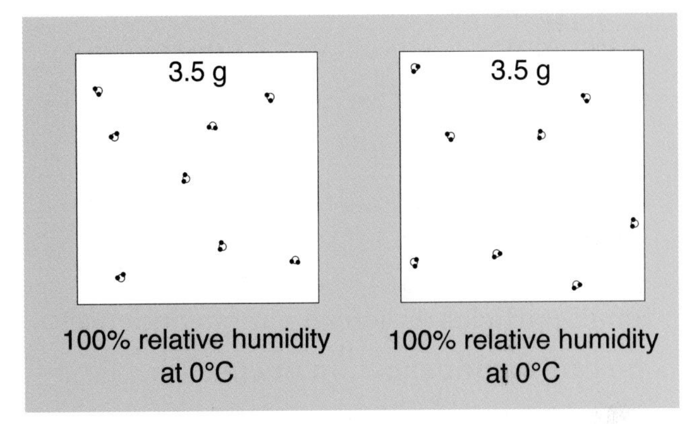

100% relative humidity at 0°C

100% relative humidity at 0°C

What happens when the air gets even colder? The table shows that a kilogram of air at –10°C can hold only 2 g of water vapor. Our kilogram has 3.5 g of water vapor. What happens to the extra 1.5 g of vapor? It condenses into ice crystals.

The temperature at which a volume of air is saturated with water vapor is known as **dew point.** When the temperature drops even a tiny bit below the dew point, water condenses as ice crystals, dew, fog, or clouds.

Condensation Surface

Water vapor needs a surface on which to condense. When the air reaches dew point, water will condense on grass, leaves, and windows as dew. Dew is a thin layer of tiny drops of water. Dew forms on large surfaces.

But what about fog and clouds? Here, too, water vapor needs a surface on which to condense. In the case of fog and clouds, the surface is microscopic. It can be as small as a piece of dust, a particle of smoke, or a tiny crystal of salt.

Small particles on which vapor condenses are called **condensation nuclei.** The air is full of tiny things that can act as the nucleus around which condensation starts. Once a droplet is started, more water vapor can condense on the surface of the water droplet.

Back to Dragon's Breath

Air in your lungs is warm—close to 35°C—and humid. In fact, the humidity of an exhaled breath is at or very near 100% relative humidity. Most of the time the water vapor in the exhaled humid air enters the atmosphere and just adds to the humidity of the air. When the warm air from your body is exhaled on a cold day, it cools rapidly. Cold air holds less water vapor. Your breath air quickly drops to dew point and becomes saturated with water vapor. The water vapor from your breath instantly condenses on invisible dust particles present in the air, and you let out a blast of cold dragon breath!

The colder the air temperature, the easier it is to saturate the air with the water vapor from your breath. So where could you go today to breathe out a cloud? Your face would have to be in the cold even if the rest of you wasn't. Open the freezer compartment on your refrigerator and blow a blast of dragon breath on the frozen peas and carrots. But be careful, you don't want to defrost the freezer by accident.

Think Questions

1. **What is relative humidity?**
2. **What is dew point? What does dew point have to do with dragon's breath?**
3. **Why does fog form on bathroom mirrors and car windows?**
4. **On what kind of day would it be possible to create frozen dragon's breath?**

Observing Clouds

Fluffy, puffy white clouds in a bright blue sky. This is one of the first memories of clouds that many of us have. You might remember a peaceful time lying on your back looking at the sky above, imagining shapes coming to life in the clouds. Ducks, people, trucks, houses, and horses might have paraded by as the puffs of bright cloud slowly changed.

You've already learned that clouds form when water vapor condenses on tiny particles of smoke, salt, and other condensation nuclei. But why do some clouds appear puffy and white and others grow to towering mountains? And what about those clouds that cover the sky as a gray, gloomy mass? Why are some clouds close to the ground and others faint streaks high above?

Clouds appear as one of two basic types—**cumuliform** and **stratiform.** *Cumuliform* describes the puffy, sometimes fast-moving and rapidly growing kind of cloud. *Cumulus* comes from the Latin word that means "heap." To grow a cumuliform cloud, air must be moving upward. As air rises, it cools. If water vapor and condensation nuclei are present, you've got the ingredients for a cumuliform cloud. When you see cumuliform clouds, you can infer that the weather conditions are unstable, and change may be in the works.

Stratiform clouds are flat and layered. *Stratus* is a Latin word meaning "layer." Stratiform clouds form when weather conditions are fairly stable. They result from the lifting of a large, moist air mass.

Meteorologists also observe where in the troposphere clouds form. High-level clouds form above 5000 meters (m) and are given the *cirro-* prefix. Middle-level clouds form between 2000 and 5000 m and are given the *alto-* prefix. Low-level clouds form below 2000 m. There is no special prefix for low-level clouds.

Some clouds may extend from low to high levels. They are nimbus clouds. *Nimbus* means "rain-bearing."

Luke Howard, the Cloud Father?

Luke Howard is sometimes called the godfather of the clouds.

Howard was never trained as a scientist, but he loved nature, especially weather, from an early age. For more than 30 years of his life, he kept a record of his weather observations. He presented his first system for classifying clouds in 1802. It is the same system meteorologists all over the world use today.

Howard also discovered that the air over cities is warmer at night than air over the countryside. We call this an urban heat island today.

You can describe just about any cloud you observe by its shape and altitude. For example,

- An altostratus cloud is a middle-level-layer cloud.
- A cirrocumulus cloud is a high-level puffy cloud.
- A cumulonimbus cloud is a heaped cloud growing from low to high levels, bringing rain.

Some low-level clouds have no prefix and are just known as stratus or cumulus clouds.

The words that describe clouds are very useful when you're recording weather observations and want to tell someone else what you have observed. Knowing why different clouds form gives you a good idea of weather conditions in your area.

That which no hand can reach, no hand can clasp.

A description of clouds in a poem
by Johann Wolfgang von Goethe (1749–1832)

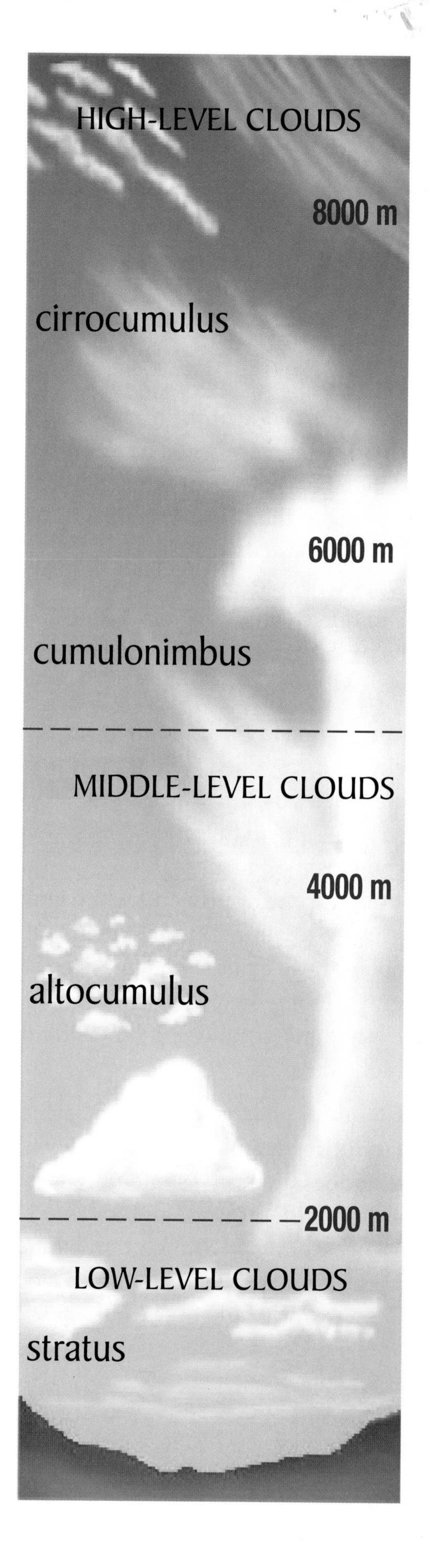

LOW-LEVEL CLOUDS

Stratus

The base of stratus clouds is often around 600 m. Stratus clouds form in stable air. They appear flat and layered, with no lumps or bumps.

STRATUS

Stratocumulus

Stratocumulus clouds form when warm, moist air mixes with drier, cooler air. When this mixture moves beneath warmer, lighter air, it starts to form rolls or waves. It looks thick. It may bring drizzle or light precipitation.

STRATOCUMULUS

Cumulus

Puffy white clouds at low levels are called cumulus clouds. When they are small and scattered, it means good weather. These are sometimes called fair-weather cumulus.

CUMULUS

Cumulonimbus

Cumulonimbus clouds form on hot summer days. The sky may start clear, with little wind. Air heated by the ground rises. Convection cells form. Warm air rises through the cell center; cooler air sinks down the sides. A cumulonimbus cloud forms. It is taller than a cumulus cloud, with a base between 300 and 1500 m. Rain starts to fall. Thunderstorms may develop.

MEDIUM-LEVEL CLOUDS

Nimbostratus

Nimbostratus clouds are sheet clouds carrying rain. Rain or snow falls almost continuously. There is usually little turbulence.

NIMBOSTRATUS

Altostratus

Altostratus clouds appear white or slightly blue. They can form a continuous sheet or look fibrous. They form between 2000 and 5000 m. Rain or snow may fall. Sometimes you can see the Sun through an altostratus cloud.

ALTOSTRATUS

Altocumulus

Altocumulus clouds form between 2500 and 5500 m. They look like small, loose cotton balls floating across the sky.

ALTOCUMULUS

Altocumulus mammatus

Altocumulus mammatus clouds look threatening, but actually indicate that the rainy weather is almost over. The clouds droop because the air is cooling and sinking.

ALTOCUMULUS MAMMATUS

Altocumulus lenticularis

Altocumulus lenticularis clouds are lens-shaped. Sometimes they look like flying saucers. They form at the top of a wave of air flowing over a mountain peak or ridge.

ALTOCUMULUS LENTICULARIS

HIGH-LEVEL CLOUDS

Cirrus

Cirrus clouds are made of falling ice crystals. The wind blows them into fine strands. The longer the strands, the stronger the wind. Cirrus clouds indicate that the air is dry. Good weather should continue.

CIRRUS

Cirrocumulus

Clouds composed of many smaller clouds at a high level are called cirrocumulus. Some people think these clouds look like fish scales. It is sometimes known as a mackerel sky. It may mean that unsettled weather is on its way.

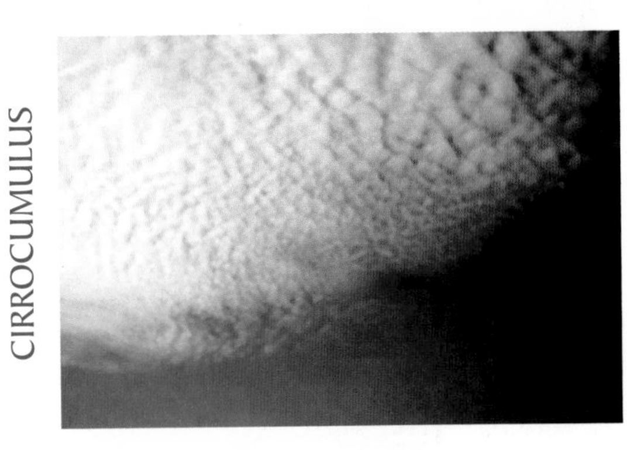
CIRROCUMULUS

Cirrostratus

Cirrostratus clouds are high-level clouds that cover the sky. The cloud is thin and transparent. You can see the Sun or the Moon through cirrostratus clouds.

CIRROSTRATUS

Jet Contrail

This jet is flying among some cirrus clouds. This could be called a human-made cloud. The jet's contrail is formed by condensation of water vapor from its exhaust.

JET CONTRAIL

Think Questions

1. Look at the weather observations your class has recorded on the class weather chart.
   - If you included cloud observations, what were the most common types of clouds?
   - Try to identify any relationships between the types of clouds and other weather observations. For example, when air pressure decreased, did a certain kind of cloud appear?
2. If stratus clouds fill the sky for several days, what does that tell you about the stability of the air? What kind of weather might you expect?
3. Cumulonimbus clouds often form in the afternoon. What weather and land conditions might contribute to their forming later in the day? (Hint: Think about solar heating of Earth and heat transfer.)
4. Read the quote by Goethe on page 38. What do you think he means?
5. Write and illustrate a short poem about clouds.

WEATHER BALLOONS AND UPPER-AIR SOUNDINGS

In the late 19th century, meteorologists used kites to gather data about the air above them. Kites could fly up to 3 kilometers into the air. Temperature, pressure, and humidity data were gathered. Kites worked well when the wind cooperated.

The **radiosonde** was developed in 1943. A radiosonde is a weather-instrument package that can be carried into the stratosphere by a balloon. It has sensors for measuring temperature, relative humidity, and air pressure. Measurements are taken continuously as the balloon rises. A radio transmitter sends the data to a ground receiver. A tracking device monitors the location of the radiosonde during its flight. Wind speed and direction are calculated from the tracking data.

A weather balloon is made of a thin membrane of natural or synthetic rubber. It is inflated with either hydrogen or helium. A biodegradable plastic parachute is attached to the radiosonde. The balloon expands as it rises. When the balloon bursts, the radiosonde is carried to Earth by the parachute.

A radiosonde can be used as many as seven times. About one-third of the radiosondes launched by the National Weather Service (NWS) are recovered. Instructions are printed on each radiosonde, explaining how to return the device to the NWS. It goes to the NWS Instrument Reconditioning Branch in Kansas City, Missouri, where it is made ready for another flight.

Weather kites were used to take recording instruments to high levels. Temperature, pressure, humidity, and winds were observed from kites.

Recording instruments traveled with the kites. They were brought back to the ground, where the observations could be read.

There are more than 900 upper-air observation stations around the world, 108 of them in the United States. Most stations are located in the Northern Hemisphere. Observations are called soundings. Soundings are taken at the same times each day, 00:00 and 12:00 UTC (Universal Time Coordinated), 365 days per year. The data are used for global and regional weather prediction, severe-storm forecasts, general aviation and maritime navigation, ground truth for satellite data, weather research, and climate-change studies.

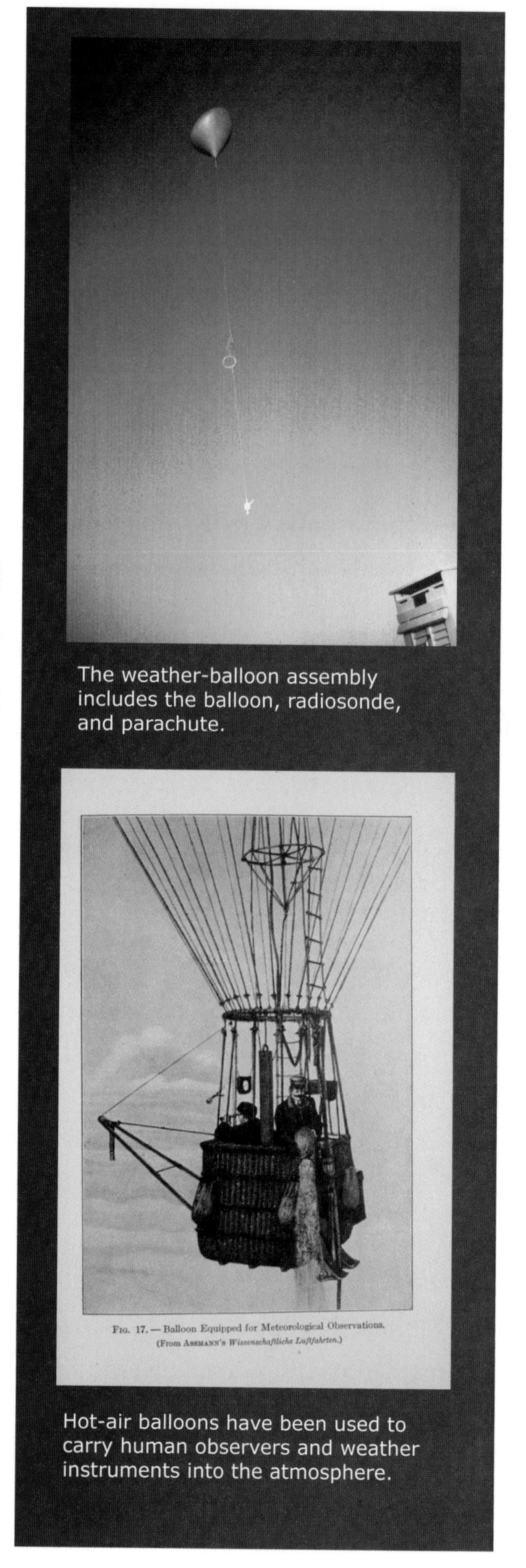

The weather-balloon assembly includes the balloon, radiosonde, and parachute.

Hot-air balloons have been used to carry human observers and weather instruments into the atmosphere.

THINK QUESTIONS

1. WHAT IS A RADIOSONDE?
2. WHEN DO METEOROLOGISTS LAUNCH WEATHER BALLOONS?
3. WHAT ARE SOME OF THE ADVANTAGES OF USING BALLOONS TO GATHER WEATHER INFORMATION?
4. WHY DO WEATHER BALLOONS EXPAND AS THEY RISE THROUGH THE TROPOSPHERE?

Earth is known as the water planet. Of all the planets in the Solar System, Earth is the only one that has vast oceans of water. If your first view of Earth from space was of the Pacific Ocean, you might think Earth was completely covered with water.

Earth is a closed system (almost). That means almost no material comes to Earth from space, and, on the flip side, no material is lost. This includes the water. All the water that is here now has been here for billions of years. The good news is that Earth's water is going to stay here for the next several billion years, but the probability of getting any more water from some outside source is essentially zero. What you see is what we've got.

Where Is Earth's Water?

By now, you know water is almost everywhere—in the oceans, in and on the land, and in the atmosphere. The pie charts show how water is distributed on Earth. A quick glance shows that just about all Earth's water is in the oceans. All that water and not a drop to drink, because to humans, seawater is poisonous. To acquire the water we need for survival and to support civilization, we need access to the small portion of Earth's water that is fresh. After subtracting the amount of fresh water frozen as icecaps and glaciers and as groundwater, that small portion is less than 1% of the water. Even this small percentage is still a pretty impressive volume of water—about 13 million cubic kilometers. This water is in lakes, rivers, swamps, soil, snow, clouds, water vapor, and organisms. It is known as free water because it is free-moving and constantly being refreshed and recycled on and over the land.

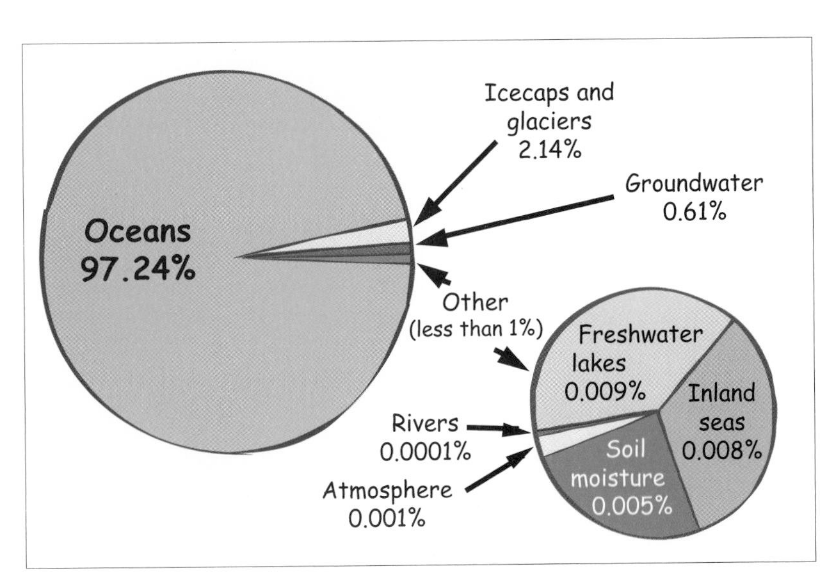

Source: *The Hydrologic Cycle* (pamphlet), U.S. Geological Survey, 1984.

Most of the water we can easily use comes from rivers and lakes. Water in rivers and lakes is known as surface water. Water that falls as precipitation can either remain as surface water or seep into the ground, where it is stored in soil or porous rock. Underground water is known as groundwater. You can see from the pie charts on the previous page that there is much more water stored underground than at the surface. It's water that is close at hand, but water that we can't see.

Water Use

Americans place high demands on water sources. Think about this. In 1995, people in the United States used about 1204 billion liters of surface water a day. They also used about 289 billion liters of ground-water a day. That's a total of nearly 1500 billion liters every day. Over the course of a year that adds up to more than 500,000 billion liters! That translates into half a million cubic kilometers per year. This is a significant percentage of the 13 million cubic kilometers of free water available on Earth—about 4%.

Flooded farm in the Midwest.

People use water in many different ways. Most important, water is essential for life. Without water to drink, we wouldn't survive. You can probably think of many nonessential ways you use water at home. You wash clothes, brush your teeth, and cook food with water. Swimming pools are filled with water, and lawns are watered. Humans also use water for navigation, for creating electricity, in manufacturing, and for agriculture. Many of these activities require good water quality. And, unfortunately, many of these activities create pollutants that can lower water quality.

Water Distribution

Water is distributed on Earth's landmasses by weather. If weather did not continually resupply the land with water in the form of rain and snow, all land would be arid and lifeless. Weather does not, however, distribute water equally around the planet. Some places, like the Midwest, have variable water supplies, getting very little precipitation one year and too much the next. During droughts there may be severe water shortages, followed by floods. The deserts of the world are always parched, while the tropical rain forests are continually soaked. Adding to the problem of water distribution is the pattern of human-population distribution. Some densely populated areas, like Los Angeles, Phoenix, and New York City, need more water than is available locally. They have to import water from faraway places.

Scientists are concerned about the warming trend on Earth. Global warming could affect both evaporation and precipitation in the United States. If more evaporation happens than precipitation, the land will dry, lake levels will drop, and rivers will run at lower levels. Other regions may receive more precipitation than usual,

Photo taken by space-shuttle astronauts of Valley of the Kings, southern Egypt, October 1988. Water from the Nile River is used to water crops. The land is arid outside of the agricultural area.

A middle school field trip to Kings Landing on the Patuxent River, Maryland. Biologists are dragging a seine net to determine kinds and variety of organisms present in the river at this time of year. Photo by Mary Hollinger, NODC biologist, NOAA.

creating floods and affecting vegetation. It will take worldwide planning and cooperation to adjust to the impact of global warming.

Earth is the water planet. Fortunately, water is one of our renewable resources. It is constantly being recycled among the atmosphere, land, and oceans. Humans can't change how much water there is. But we can make smart decisions about how much of it we remove from natural systems, how it is distributed, how it is used, and what happens to it after we use it. As the demand for water increases worldwide due to population increase and a rising standard of living (which requires water), everyone will have to participate in water conservation. Industries will need to be more efficient with water use and careful not to introduce pollutants into water sources. Agriculture will have to develop more conservative crop-watering practices. And every citizen will have to become more aware of the value of water and treat it as the most precious substance on Earth.

References

Environmental Protection Agency/Water, Resources Site

www.epa.gov/globalwarming/impacts/water/index.html

U.S. Geological Survey, Water Science for Schools

ga.water.usgs.gov/edu/

America's Water Supply

www.gcrio.org/CONSEQUENCES/spring95/Water.html

What Is Air Pressure?

The atmosphere is composed of air. Air has mass. In fact, a column of air 1 centimeter square (cm^2) extending up to the top of the atmosphere has a mass of 1.2 kilograms (kg). If the top of your head has a surface area of 150 cm^2, that means every time you go out and stand under the open sky, you have the pressure of 180 kg of air pushing down on your head! That's like wearing a hat with a refrigerator on it. Is it safe to go outside?

1 cm^2

Don't worry, it's safe. The force applied by the **air** above you is called **atmospheric pressure.** Life on Earth has evolved in this high-pressure environment, so we are able to handle the pressure just fine. In fact, most of the time we are totally unaware of the pressure, except...

We do feel the air pressure when it changes quickly. Have you ever traveled to the mountains and noticed an interesting sensation in your ears as you go up or down the mountain? It's called popping your ears. Sometimes this happens in an airplane when it changes altitude rapidly or in a car going up and down a mountain road. If you have had that experience, keep it in mind as you think more about atmospheric pressure.

Sometimes you can see evidence of change in pressure even if you can't feel it. In an airplane, you might notice that packets of peanuts or chips are puffed up like balloons.

Or if you drain a plastic water bottle up in the mountains and screw the cap on tightly, it might get squashed a little as you drive down the mountain. These are examples of air pressure.

WHAT CAUSES AIR PRESSURE?

Molecules have mass, so they are pulled to Earth by gravity. The air surrounding Earth has weight. Atmospheric pressure is the weight of the air pushing on Earth's surface.

Remember, air molecules are zipping around individually. So what prevents gravity from attracting them all to the ground? Why aren't we walking around knee deep in a soup of oxygen and nitrogen molecules?

The answer is kinetic energy. The gas molecules have so much energy of motion that they are pushing each other away in all directions. They resist being crowded together by this constant banging into one another.

Here's an important point: The pressure is not only pushing down on Earth; the molecular banging is also pushing back with equal force on the air molecules above. Also, molecules are pushing back with equal force on the molecules trying to crowd in from the sides. Atmospheric pressure acts with equal force in every direction.

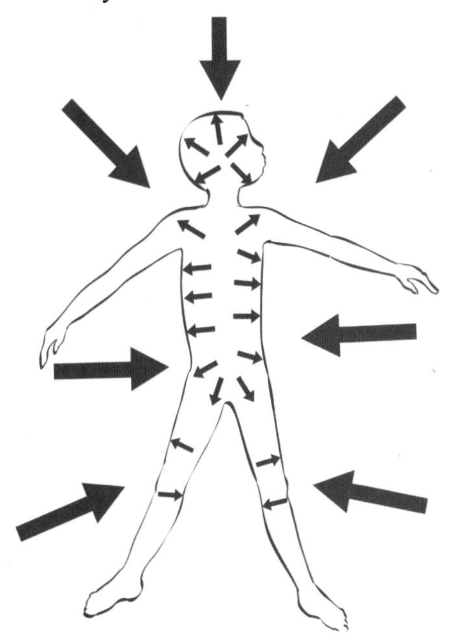

Atmospheric pressure is not the same everywhere. Air pressure is caused by the mass of air being pulled to Earth. But what happens if you go up into the atmosphere, high above Earth's surface?

If you have the good luck to go for a hot-air-balloon ride, you might find yourself 2 kilometers (km) above the land. Up there, 2 km of air is below you, so that 2 km of air is not applying pressure up where you are. The atmospheric pressure is less up in the balloon.

The greater the amount of air overhead, the greater the pressure. Also, the greater the pressure, the closer together the molecules are pushed. Remember, gases can be compressed. Pressure (force) drives gas molecules closer together. When more molecules are present in a given volume, the gas is denser. Because we live at the bottom of a sea of compressible air, the atmosphere is densest at Earth's surface. It becomes less and less dense as you go higher in the atmosphere.

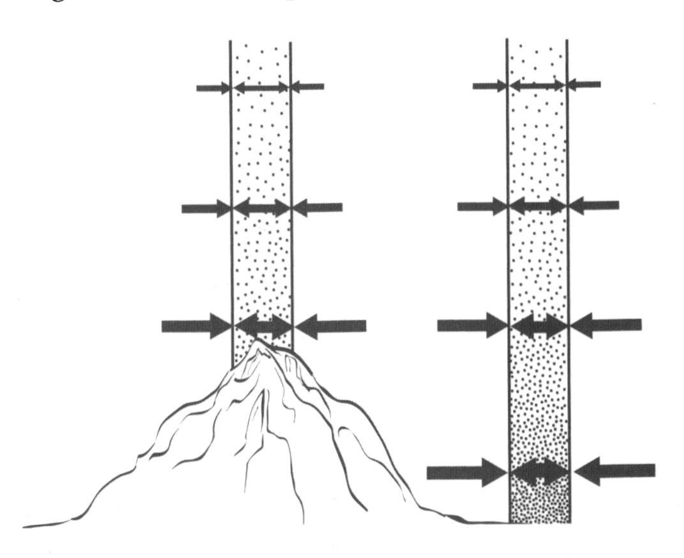

As you go up in the atmosphere, pressure goes down. As pressure goes down, the air expands (less force pushes the molecules together). As air expands, it gets less dense.

Mount Everest is over 8 km high. Up on top the atmospheric pressure is only one-third the pressure at sea level. Consequently, the air is one-third the density of air at sea level. Have you ever seen pictures of climbers laboring up the highest reaches of the mountain? Most of them are using oxygen supplies. Why? Because there is only one-third as much oxygen in each breath of air at that elevation. It takes an exceptional climber to reach the summit without extra oxygen.

Now, about those ears popping when you travel through the mountains. Does it have anything to do with atmospheric pressure?

MEASURING AIR PRESSURE

The air pressure that meteorologists talk about on the evening news is the pressure exerted by the mass of air pushing down on a certain point on Earth's surface. Elevation is one factor that causes pressure to vary, but there are a number of other factors as well. These factors, and the resulting pressure, are of interest to weather forecasters.

Meteorologists use a **barometer** to measure air pressure. An Italian naturalist named Evangelista Torricelli invented the first barometer in 1643. He filled a long glass tube with mercury and turned it upside down in a dish also filled with mercury. A small amount of the mercury (not all of it) ran out of the tube and into the dish, leaving an empty space above the mercury.

This space was a vacuum. A vacuum is a space containing almost no matter, not even air.

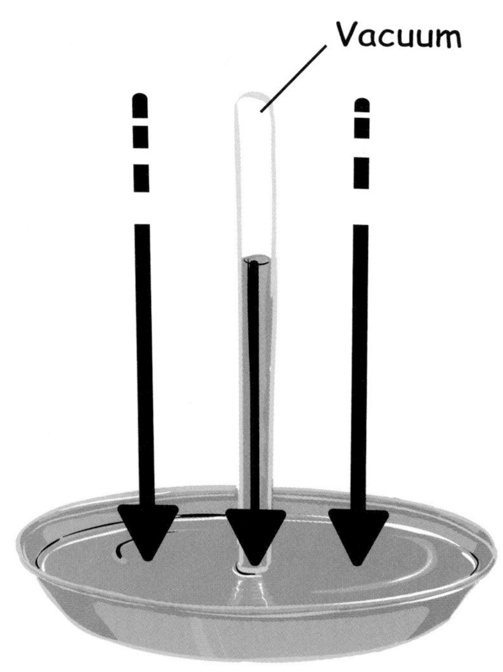

What was holding the heavy column of mercury up in the tube? Atmospheric pressure. Air pressure pushes down on the mercury in the dish. The pressure is distributed throughout the mercury, including the mercury in the tube. Remember, a column of atmosphere 1 cm^2 has a mass of 1.2 kg. If the column of mercury is 1 cm^2 in cross section, it will have a mass of...that's right, 1.2 kg. So, the air pressure is exactly balanced by the mercury pressure.

Because mercury is very dense, a column of mercury exactly 76 cm high will balance a column of atmosphere 600 km high at sea level. As Torricelli observed his new invention closely, he noted that the level of mercury moved up and down a little from day to day. He reasoned that the changing level of mercury was due to changes in the atmospheric pressure. Torricelli had invented the first barometer—an instrument for observing and measuring changes in atmospheric pressure.

Today, meteorologists often use another type of barometer called an **aneroid barometer.** Aneroid barometers are much smaller and more versatile than mercury barometers.

At the heart of an aneroid barometer is a sealed bellowslike chamber with a spring inside. All the air is removed from inside the bellows. Air pressure tries to squash the bellows flat, but the spring inside pushes back to keep that from happening. The force of atmospheric pressure and the force exerted by the spring are balanced.

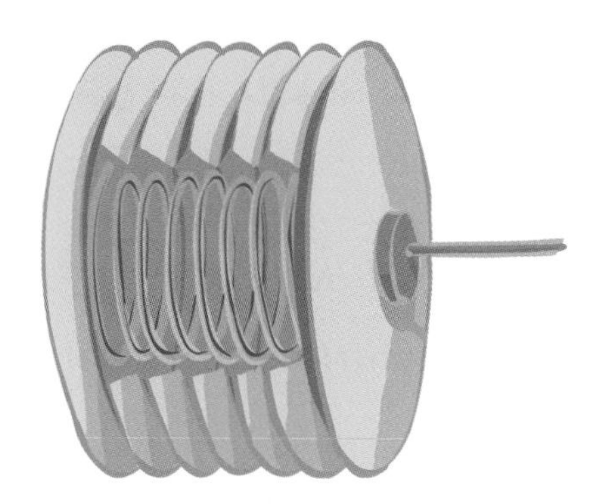

If the atmospheric pressure increases, it will push on the bellows. The ends of the bellows will be pushed closer together until the force pushing back by the spring is equal to the increased air pressure. A pointer attached to the bellows moves along a scale to show the change in pressure.

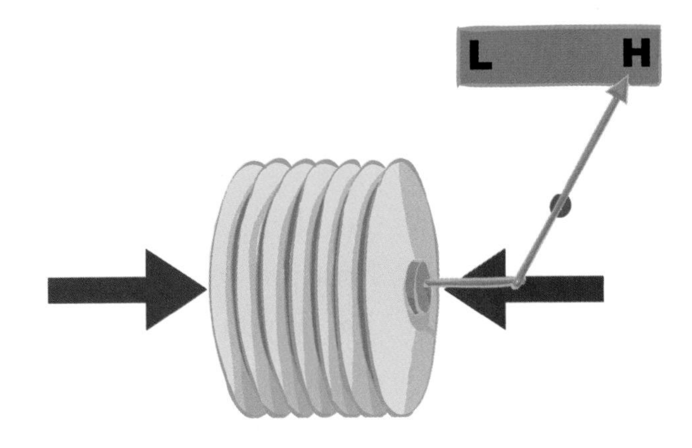

Lower pressure allows the spring inside the bellows to push the ends of the bellows farther apart.

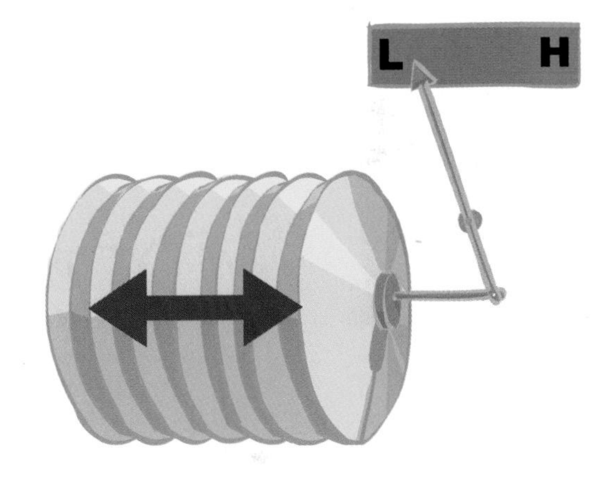

Sometimes a pen is attached to the bellows. The pen records the air pressure on a rotating cylinder to obtain a continuous record of pressure changes. If an electronic sensor and a transmitter are attached to a barometer, pressure information can be radioed from a weather balloon back to a receiving unit on Earth.

Because pressure changes with elevation, airplanes use a type of barometer to monitor how high the plane is. This application of a barometer is called an **altimeter.**

Scientists and meteorologists use several different units to report pressure. For historical reasons, inches or centimeters of mercury are used in science experiments. Meteorologists, on the other hand, look at the average atmospheric pressure at sea level and call that 1 bar. If you are relaxing at the beach (sea level), the pressure around you will be 1 bar (or close to it).

The bar has been subdivided into 1000 equal parts called **millibars (mb).** Standard atmospheric pressure is 1000 mb. These are the units you used to record your local atmospheric pressure on the class weather chart.

In practice, standard pressure is actually 1013 mb. Any pressure below 1013 mb is lower than normal pressure, and over 1013 mb is higher than normal.

Think Questions

1. *When you drive down a mountain, what makes your ears experience those interesting and sometimes uncomfortable sensations?*

2. *Why doesn't air pressure crush an empty soda can sitting on a table?*

3. *If a meteorologist says that the air pressure is getting lower, what would you expect to see happen to Torricelli's mercury barometer?*

4. *If Torricelli had drilled a little hole at the top of the glass tube holding his mercury column, what would have happened to his barometer?*

Where the Wild Wind Blows

All renewable energy (except tidal and geothermal) comes from the Sun. The Sun radiates 100,000,000,000,000 kilowatts of energy to Earth every hour. About 1–2% of that energy is converted into wind energy.

The region near the equator, extending about 20° north and south, is known as the tropics. The Sun's rays hit Earth with the greatest intensity here. The tropics, which are predominantly oceanic, absorb a lot of heat.

As the air rises, it cools. At about 10 kilometers (km) altitude, the warm air has cooled to the same temperature as the surrounding air. The cool air begins to fall back to Earth. But, because the wall of warm air is rising from the tropics, it can't return directly back. The cool air, still at about 10 km, flows north and south, like two gigantic sheets. When it reaches about 30° north and 30° south, it descends back toward Earth.

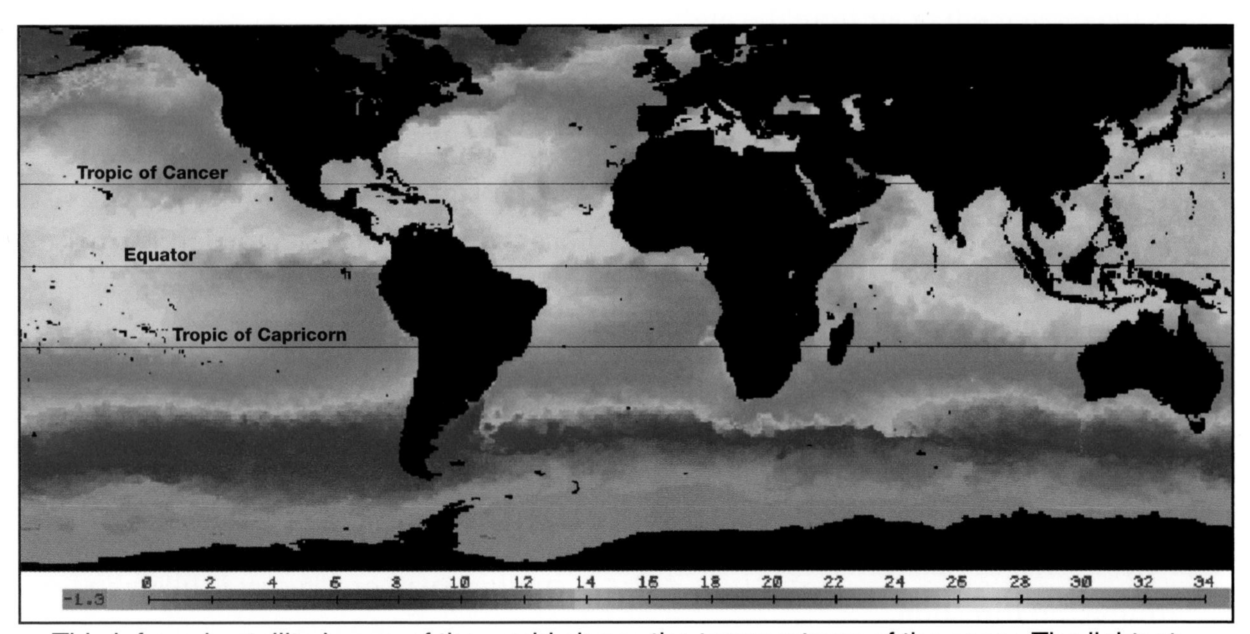

This infrared satellite image of the world shows the temperatures of the seas. The lightest areas are the warmest.

Warm tropical water transfers energy to the air. The air warms and expands. Hot air is less dense and rises into the atmosphere. Warm air does not rise in only one place, like smoke from a fire. Air rises like the smoke from thousands of fires all the way around the world in the tropics.

Meanwhile, the warm, low-density, rising air in the tropics creates a low-pressure area. Cooler, denser air that is descending from the upper atmosphere flows into the area of low pressure to replace the rising air. This creates a huge **convection cell.** The bottom of the cell flows across the surface of the planet, from about 30° north and south to the tropics, producing wind.

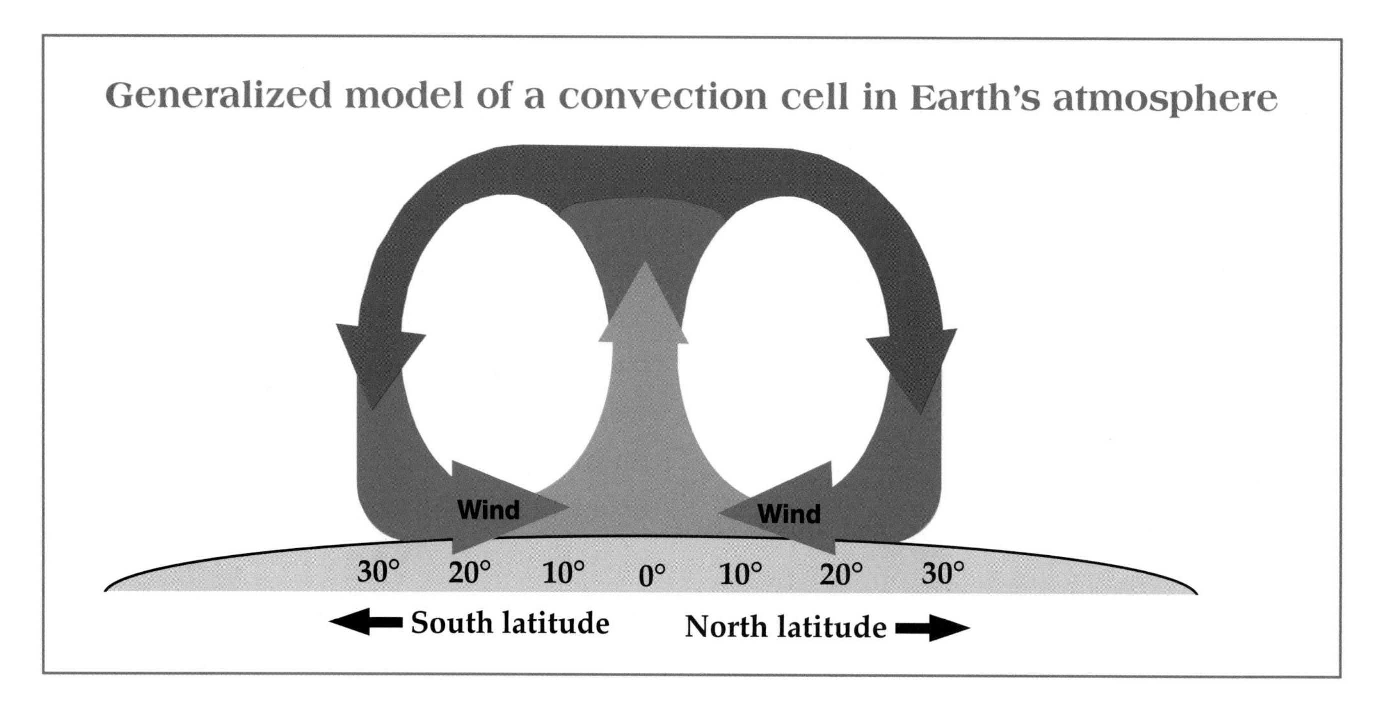

Wind is the movement of air from the high-pressure region north and south of the tropics rushing to balance the low-pressure region created by air rising from the tropics.

Differential heating creates high- and low-pressure areas, creating wind. Winds always move from a high-pressure area to a low-pressure area. Earth rotates under the moving atmosphere, and this easterly movement keeps the winds from going straight north or south, bending them to the east or west. This Coriolis effect gives us the **prevailing wind direction.**

Prevailing winds are global winds. They are predictable for different latitudes on Earth. The prevailing wind directions are important to people sailing ships, flying balloons around Earth, and building windmill farms to generate electricity. These global winds are not greatly influenced by structures on Earth's surface, like forests, cities, and mountains.

Local Winds

Local winds change with the season and even with the time of day. They are the direct result of local differential heating. Local winds are affected by land structures and bodies of water near landmasses.

A typical Chicago weather forecast in the summertime might go something like this.

Sunny skies today with a high of 85°F inland; temperatures in the low 70s lakeside.

A similar weather forecast is likely in Los Angeles and San Diego, with temperatures being cooler at the beaches than inland.

The climates of Los Angeles and Chicago are affected by large bodies of water. Landmasses near oceans or the Great Lakes will feel the effects of the nearby water. Here's how.

Landmasses get hotter faster than water when the Sun shines. The hot land transfers heat to the air above it, and the air expands.

The warmer, less-dense air creates a low-pressure area over the land.

Even though the sea also absorbs energy, its temperature does not change much at all. The sea stays cooler than the land. Less energy transfers to the air over the water, so it is cooler and denser. The air pressure over the water is higher than it is over the land.

Pressure always tries to equalize, so air from the high-pressure area over the water flows into the area of low pressure. Wind blows from the sea onto the land. This is called a **sea breeze.**

At sunset, there may be a period of calm when land and sea temperatures are about equal. After sunset, the land cools off quickly. The air over the land cools and contracts, becoming denser. Now the local high-pressure area is over the land. Meanwhile, the sea temperature is still about the same as it was during the day, so the air pressure is about the same. Even though the pressure is the same, it is lower than the local pressure over the land. Now the local high pressure over the land moves into the lower-pressure area over the water. Wind blows from the land out to sea. This is called a **land breeze.**

Measuring the Wind

A lot of people want to know how fast the wind is blowing—hot-air balloonists, airplane pilots, sailors, hang-glider pilots, and kite fliers, to mention a few. Before ships were powered by steam, most international maritime travel, trade, and war was conducted in ships driven by wind. Understanding wind was a critically important part of life on the sea.

In 1805, Sir Francis Beaufort, commander of a British naval ship, developed a simple scale for comparing winds. Beaufort was born in Ireland in 1744, and he entered the Royal Navy when he was only 13!

Beaufort's scale was a wind *force* scale, not a wind *speed* scale. Beaufort described 13 levels of wind force that he could recognize from the deck of his ship. His descriptions related to how a ship was pushed or hindered by the wind, how many sails could be flown, how full the sails would be, and, at the extremes of his scale, the chances of survival. By 1838, the Beaufort scale was the mandatory system for reporting wind force in the official log in all ships of the British Royal Navy.

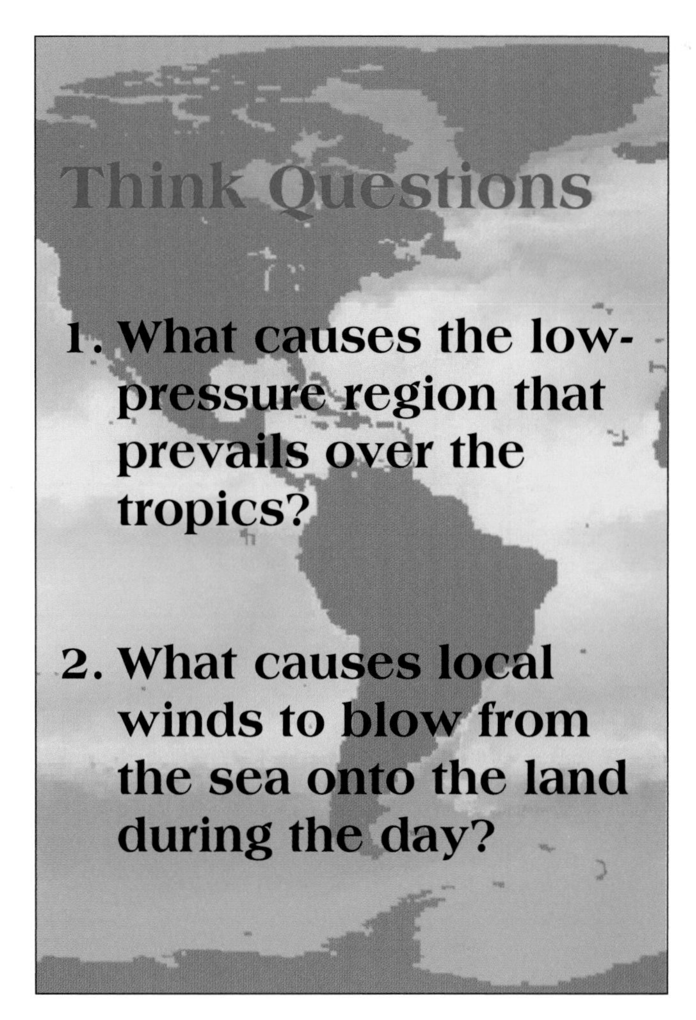

The Beaufort Scale

Beaufort number	Wind (km/h)	Wind (mph)	Wind classification	Wind effects on land	Wind effects on water
0	1	<1	Calm	Smoke rises vertically	Water calm, mirrorlike
1	1–5	1–3	Light air	Smoke drift indicates wind direction; still wind vanes	Scalelike ripples with no foam crests
2	6–11	4–7	Light breeze	Leaves rustle; wind felt on face; wind vanes moved by wind	Small wavelets; crests have a glassy appearance and do not break
3	12–19	8–12	Gentle breeze	Leaves and twigs constantly moving; light flags extended	Large wavelets; crests begin to break; scattered whitecaps
4	20–29	13–18	Moderate breeze	Dust and loose paper raised; small branches move	Small waves 1–4′ becoming longer; many whitecaps
5	30–38	19–24	Fresh breeze	Small trees with leaves begin to sway	Moderate, longer waves 4–8′; whitecaps common; some spray
6	39–50	25–31	Strong breeze	Larger tree branches moving; phone lines whistle	Larger waves 8–13′; whitecaps common; more spray
7	51–61	32–38	Near gale	Whole trees moving; difficult to walk against wind	Sea heaps up; waves 13–20′; crests break; white foam streaking off breakers
8	62–74	39–46	Gale	Twigs break off trees; difficult to walk against wind	Moderately high waves (13–20′) with greater lengths; crests beginning to break into foam that is blown in white streaks
9	75–86	47–54	Strong gale	Slight damage to buildings; shingles and slates torn off roofs	High waves of 20′; rolling seas; dense streaks of foam; spray may reduce visibility
10	87–101	55–63	Storm	Trees uprooted; considerable structural damage to buildings	Very high waves (20–30′) with overhanging crests; sea white with blown foam
11	102–115	64–72	Violent storm	Widespread damage	Huge waves (30–45′); foam patches cover sea; air filled with spray; visibility reduced
12	> 115	> 72	Hurricane	Widespread damage	Huge waves (over 45′); air filled with foam; sea completely white with driving spray; little visibility

LAURA'S BIG DAY

Laura opened her eyes with a start. She checked her alarm clock. Only 6:30 a.m. Laura got out of bed and went to the window. All she could see was the dim outline of the tree in the front yard. Fog blocked out everything else on her street.

It was her birthday. How could the weather be so lousy? Laura really wanted the fog to go away. After all, she had something very special to do today. Uncle Ken was taking her hang gliding for her 13th birthday.

They had been planning the event for 6 months, about the time Laura reached 5 feet tall. Ever since she could remember, Laura had wanted to fly like a bird. She had stood on the tops of rocks with her arms stretched out, while her parents called to her to come down. She had flown in an airplane when she visited her grandparents, but that just wasn't the same. Laura peered into the fog, hoping to see the mountains beyond.

Laura was dejected and wide awake. When she went into the kitchen, she found her mom making coffee. "The weather is lousy!" Laura pouted.

"It will probably clear up by noon," her mother offered. "I saw the weather forecast last night. The meteorologist said that it was really hot out in the desert yesterday. A high-pressure area pushed ocean air over our town all day yesterday. When the humid sea cools off at night, the water vapor condenses, forming fog. I think the Sun will come up and warm the ground and the air. The fog will evaporate again, leaving us with a nice day."

Laura wasn't convinced. She ate her cereal and stared out of the window.

"I'm going to the nursery later. Want to come along?" her mother asked.

Laura was tempted, but decided she would wait to see what the weather would do. "No, thanks," she replied.

At 9:00 it started getting lighter. Laura called Uncle Ken on the phone.

"Hi, Laura," Uncle Ken said. "You looking forward to your lesson today?"

Laura could hardly believe her ears. "Are we really going to be able to fly?" she blurted out.

"Sure," Uncle Ken said. "It's really gorgeous up here on the mountain, and I can see that the fog down your way is burning off."

"What do you mean, burning off?" Laura asked.

"The Sun just warms up the air and the fog evaporates," Uncle Ken replied.

"Oh," said Laura, "that's just what Mom said."

"Get your mom to drive you up here and we'll get the equipment ready. I'm already working on our flight plan."

Uncle Ken lived in a cabin on a mountain ranch where he worked when he wasn't teaching hang gliding or building new equipment. When Laura arrived at the cabin, she trotted down to the storage garage where Uncle Ken was working.

"What are we going to do?" she asked.

"Well, your first lesson will be a tandem flight so I can show you what hang gliding is all about," Uncle Ken answered. "We'll fly together on a glider like this Falcon here in this picture. Right now it's packaged for the road on top of my truck. I have to check on the weather now."

Uncle Ken had taken Laura with him in the spring when he taught hang gliding. She knew the names of the parts of the glider and had tried to balance one on her shoulders. She had listened to Uncle Ken talk about mature decisions and safety. Laura already knew about helmets, reserve parachutes, and harnesses. She had been at the landing zone with Uncle Ken, watching his friends land their gliders after long flights riding the thermals. She saw them look at the wind sock for wind direction, so they could land into the wind.

Uncle Ken had explained that the summer was a good time for thermals. A thermal is the name given to a bubble or column of warm, rising air, common on sunny days. The Sun heats the ground and the air near

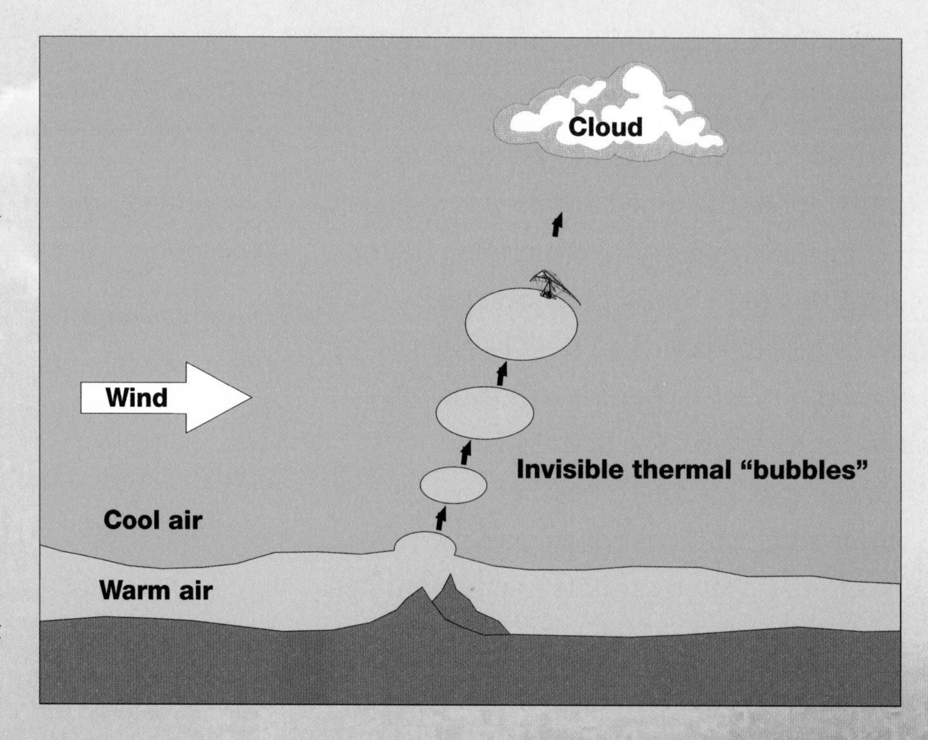

the ground. The air expands, making it less dense than the air above it. The warm air rises like bubbles rising in a glass of soda. An important skill in flying a hang glider is to spend as much time as possible in the strongest part of the thermal (usually the center).

Laura had seen Uncle Ken's weather station behind his cabin and how it was hooked up to an old computer. He recorded wind speed, wind direction, temperature, humidity, and barometric pressure. She also knew that he contacted a weather station near the landing zone to get current temperature and wind speed.

Uncle Ken used the temperature difference between his cabin on the mountain and the landing zone to figure out how strong the thermals would be. As you go higher in the atmosphere, the air cools off. It cools off at a rate of about 6°F per 1000 feet. Because the difference from the top to the bottom of the mountain was almost 5000 feet, the air should cool about 30°F. If the difference in temperature was more than that, it meant good thermals, because the warmer air at the bottom rises more quickly through the cooler air near the top of the mountain.

"Conditions look great. Let's go."

Laura tossed her duffle into the back of Uncle Ken's truck. She and her mom jumped in, and they were off.

On the road Laura asked, "Uncle Ken, how long will we be up in the air?"

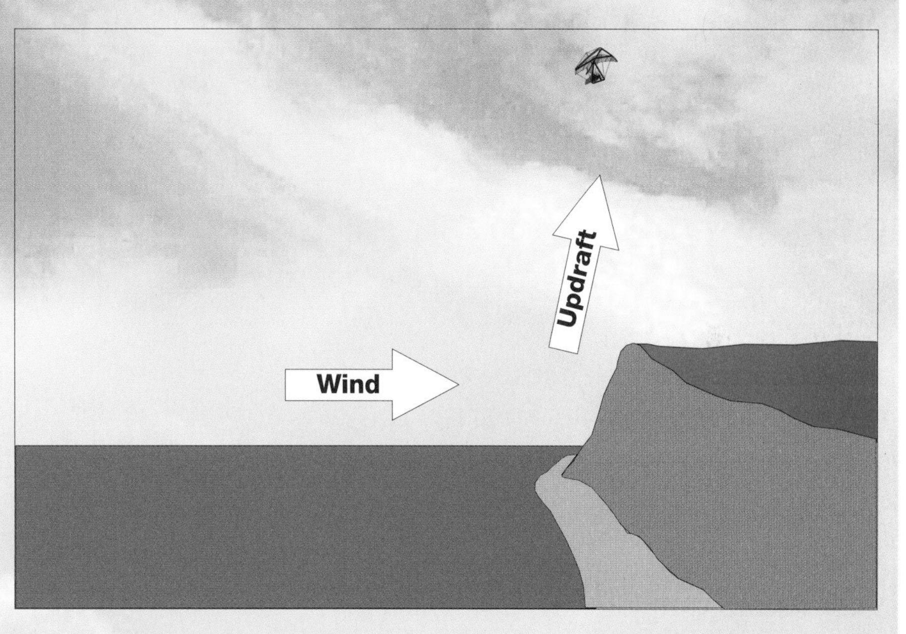

"No telling," he responded. "I was up over 2 hours on Wednesday, but I don't think we'll push it that far. Maybe we'll fly for 15 or 20 minutes. But it all depends on the wind."

Uncle Ken talked about how this site had both ridge and thermal lifts, and that promised a great ride. "Ridge lift means that the wind gets forced straight up when it runs into a cliff or hill. Hang gliders don't have motors to gain altitude, you know, so the only energy we use for going higher in the air is an updraft—currents going upward. Any time that a steady wind blows directly against a wide slope, ridge lift will probably result."

They pulled into the launch site and parked near several other trucks and vans. Uncle Ken pointed to a wind sock about half extended by the gentle, steady breeze. "It's a little old-fashioned, but we like it. It's a pretty straightforward tool for estimating wind direction and speed. Looks like we have a breeze of about 15 miles per hour."

Uncle Ken set about assembling the glider. It was surprisingly big, but he handled the lightweight materials with ease. Laura watched with interest, but she could feel a little anxiety rising in the pit of her stomach as the assembly neared completion. The moment when they would launch into thin air was drawing uncomfortably close.

Laura zipped up her warm parka and strapped on her harness.

"Is it tight enough? Does it feel right?" asked Laura's mother.

"Mom, please, I'm all right. Uncle Ken showed me how...I can do it by myself."

Uncle Ken checked and double-checked Laura's preparation. She was dressed in warm, rugged clothes, helmet, goggles, and a properly tightened harness. Ready.

Uncle Ken and Laura reviewed the procedure for getting into the air safely, the flying position, and the landing procedure. Uncle Ken could see how excited she was, and he reminded her that this was a lesson, not a ride, and that she had to be mature and follow his instructions in the air.

Uncle Ken hooked onto the hang glider and carried it to the launch area. Following his instructions exactly, Laura hooked her harness onto the glider right next to her uncle. She practiced lying out prone—just like she would in the air.

"Ready?" asked Uncle Ken.

"I think so. It's a little scary, but I want to see what it is like to fly."

Laura stood next to Uncle Ken on the edge of the gentle slope. The breeze was blowing in their faces. Laura felt the hang glider lift up as it caught the wind.

"Now! Run!"

Down the slope they dashed. Amazingly, in just a moment Laura's feet barely touched the ground. She felt as light as a leaf on the autumn wind. Then she wasn't touching the ground at all. She was in the air, and the ground was moving away with alarming speed.

"Uncle Ken, it worked! We're flying!"

Uncle Ken shifted his weight to bank the glider so they swooped over the launch area and Laura's mother. Laura waved at the amazed figure of her mother with her hands clamped over her mouth. Then her mother waved back and remembered to snap a few pictures.

"Let's use this nice ridge lift to climb a little, and then let's look for a thermal." They flew back and forth along the ridge twice, and then Uncle Ken turned away from the ridge to look for thermals. It was so quiet and peaceful. The hang glider just floated. As they headed over the valley, the land fell away and Laura realized she was thousands of feet above the ground. This was real flying.

"Look, there's a thermal just a quarter mile to the left," announced Uncle Ken after a couple of minutes. "Can you see it?"

She looked where he indicated, but could see nothing.

"Those two red-tailed hawks, spiraling up, have pointed out the invisible thermal for us. Let's join them."

In a minute, Laura could feel the turbulence in the air and the definite lift provided by the thermal. Uncle Ken skillfully banked the glider to keep it in the thermal, and they spiraled up, just like the hawks above them.

In a few minutes, the thermal lift died out, and they glided off toward the site where they would land, making large, lazy circles in the sky. When they passed through a region of turbulence, Laura asked with wide eyes, "Uncle Ken, was that a thermal? Could it take us higher again?"

Pleased by her close observation, Uncle Ken responded, "I'm not sure. Let's go see."

He banked sharply and entered the thermal. The glider was once again lifted several hundred feet in a minute.

"Excellent lift. That was a nice little elevator you spotted. Let's head for home."

Uncle Ken turned the glider away from the mountain and began to fly a circular route that took him lower and closer to the landing area. As they got closer to the landing zone, Uncle Ken asked Laura to watch the wind sock and the streamers to see which way the wind was blowing. They slowly glided lower and lower until the ground was just below them. Uncle Ken pushed the down tube forward, and the glider made an easy dive toward the landing target. They skimmed just above the ground. Uncle Ken pushed out hard on the down tube. The front of the glider rose up and they settled easily to the ground.

It was over. They were back on the ground. She had flown. Laura was dazed. "Uncle Ken, that was great! I never knew anything could be so exciting. Can we go again?"

"Well, the first hang-gliding trip is a ride. If you go up again, you will have to do some of the driving, you know."

"Can I? Can I really do the flying?"

Laura saw her father coming across the field as she and Uncle Ken carried the marvelous flying machine to the parking lot.

"Dad, I was flying and it was the greatest thing. It's like flying a kite, except we were on the kite! And you won't believe it, Dad, we followed red-tailed hawks. We flew with red-tailed hawks right to the top of a thermal!"

Laura couldn't get the idea of flying with birds out of her mind. She knew that she would continue her lessons and become a pilot. She'd have her own glider one day. But before that, there was a lot to learn about seeing, feeling, and imagining what's going on in the air, and learning about the weather, and what makes a good day for hang gliding.

Think Questions

1. **Why is the weather forecast so important for hang gliders?**
2. **How does a hang-glider pilot rise higher in the atmosphere?**

IS EARTH GETTING WARMER?

Global warming is in the news and on the scientific agenda worldwide. And there are a lot of questions about it. Is it real? Is it happening fast? Is it serious? Is it natural or caused by humans? What can we do to stop it? What will happen if we don't stop it?

The answer to the first question is yes, global warming is real. The scientific community of climatologists has evidence that Earth is heating up. Here is a sampling of what has been confirmed.

Atmosphere temperature. Earth's lower atmosphere (troposphere) has warmed by about 0.6°C over the past 100 years. This may not seem like a lot, but it is significant, and some weather effects result from even slight temperature changes.

Glaciers. A glacier is a large sheet of ice that moves very slowly. Many glaciers in the world are getting smaller. Glaciers are melting in Montana's Glacier National Park. At the present rate of melting, they will be gone in 50 years. Large chunks of ice are breaking off the Antarctic ice sheets. There isn't as much ice at the North Pole. Ice melting could be the result of global warming.

Ocean temperature. The temperature of the upper ocean layer (top 300 meters) has increased by more than 0.3°C in the past 40 years. As we know, a small increase in water temperature represents a lot of heat energy. Increased oceanic temperature can influence global weather patterns and hasten the melting of oceanic ice.

Sea level. The level of the sea is rising the world over. Over the last 100 years, the level of the sea has risen 15–21 centimeters. When the sea level rises, the tide goes farther up the beach and submerges low coastal areas. Scientists think the sea has risen because there is more liquid water in the world today. This water came from melted glaciers and sea ice. Also, heat makes water expand. When the ocean expands, it takes up more space.

What Is Causing Earth to Warm?

Earth's atmosphere produces what is often popularly called the greenhouse effect. The name suggests that Earth operates like the little glass houses used by gardeners to grow plants in cold climates. Greenhouses work by trapping heat from the Sun. The glass panels of the greenhouse let in light. The light is absorbed by plants, soil, and water in the greenhouse and reradiated as infrared rays. Infrared is also known as heat rays. The glass doesn't let the infrared escape, so the heat builds up inside the greenhouse.

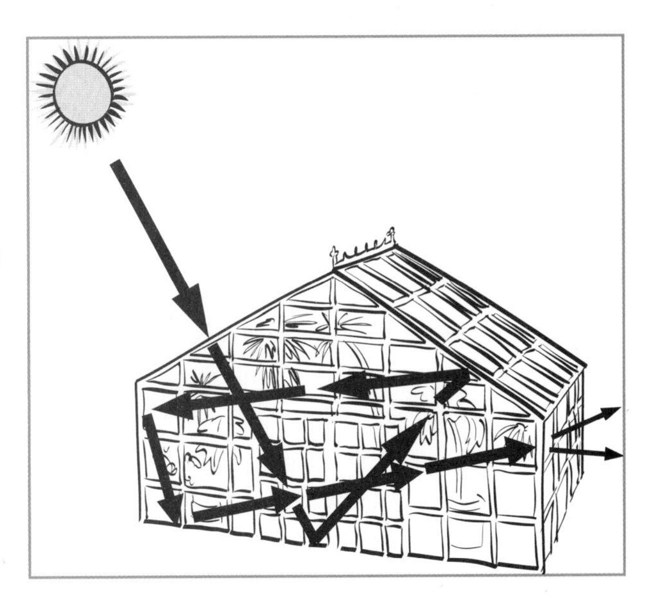

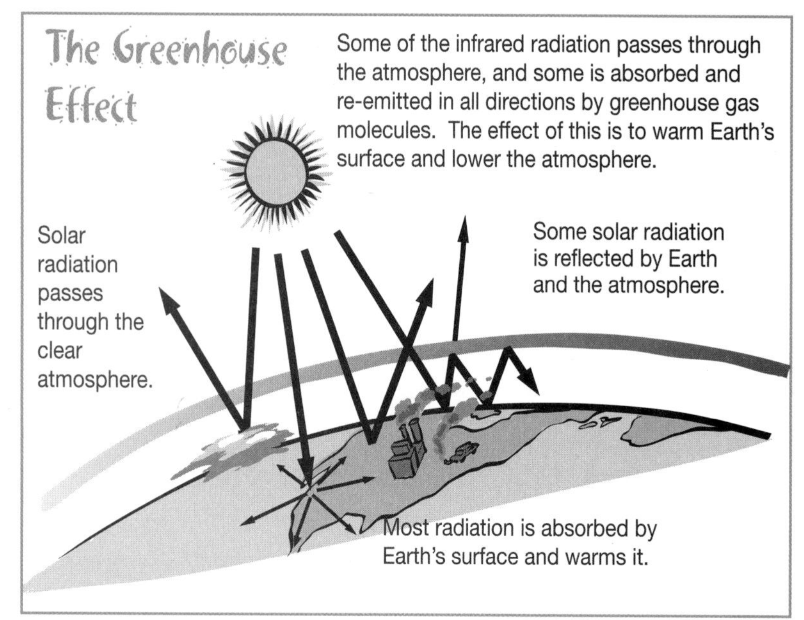

Earth is not surrounded by glass, but it is wrapped in atmosphere. Two of the important gases in the atmosphere are water vapor (H_2O) and carbon dioxide (CO_2). Because these two gases can absorb solar energy, they are called greenhouse gases. The action is not exactly like a greenhouse, because the atmospheric gases don't reflect heat back to Earth. They absorb heat, but the effect is the same. The air heats up.

CO_2 is a by-product of combustion. Whenever someone starts up a car, lights a barbecue grill, burns a field, or takes off in a plane, CO_2 is released into the atmosphere. The amount of CO_2 in the atmosphere has jumped in the last 100 years. Not coincidentally, the increase of CO_2 corresponds with the explosive use of fossil fuels worldwide and the massive burning of the rain forests.

The greenhouse effect is not all bad, however. The protective blanket of greenhouse gases prevents Earth from losing too much heat to space. Without these gases, Earth's average temperature would be about 33°C colder. The problem that needs attention is excessive greenhouse gases that may contribute to global warming.

What Do We Know about the History of Global Climate Change?

Climate is the long-term average of a region's weather patterns. Earth's climate has been changing constantly over its 4.5-billion-year history. The geological evidence indicates that during some periods the climate was so warm that sea level was 20 m higher than it is today. This would suggest that no water was locked up in ice. At other times, Earth cooled, water was trapped as ice, and sea level dropped as much as 122 m. Each of the changes may seem extreme, but they probably occurred slowly over many thousands of years.

The first people arrived in America between 15,000 and 30,000 years ago. During that time, much of North America was covered by great ice sheets called continental glaciers. Sea level was so low that people could walk to North America on a land bridge between Asia and Alaska. Some 14,000 years ago, the vast ice sheet began to melt very quickly. By 7000 years ago, the ice was gone.

In the 14th century, Europeans lived through what is known as the Little Ice Age. The Little Ice Age lasted for several hundred years. During it, glaciers advanced down mountain valleys, and hard winters and famines caused great hardship for people.

Clearly, humans didn't influence these climate changes. They occurred naturally, probably as a result of variations in energy output of the Sun. But the Sun's energy output has been steady for the last hundred years, and still the heat is building up.

What's in the Future?

Scientists are not fortune-tellers, so they don't claim to know exactly what will happen in the future. But they can use computers to model some of the possible outcomes. The computer models tell us that Earth may continue to get warmer. The information suggests that Earth's temperature will probably continue to rise as long as the concentration of greenhouse gases in the atmosphere continues to increase.

The key to halting a runaway greenhouse effect is reversing greenhouse gas emissions. There are two ways to do that: stop adding new greenhouse gases to the atmosphere, and take measures to remove greenhouse gases currently in the atmosphere. Neither is easy, but steps must be taken.

One of the most powerful ways to reduce greenhouse gas emissions is the action of individuals. The first action is understanding when you are contributing to the greenhouse load. It's obvious that CO_2 is being released when you start a gasoline engine or burn a fire. But you also contribute to CO_2 emission when you

- Watch TV
- Use the air conditioner
- Turn on a light
- Use a hair dryer
- Ride in a car
- Play a video game
- Listen to a stereo
- Wash or dry clothes
- Use a dishwasher
- Microwave a meal

We need energy to do things like drive a car, fly a plane, or make things in factories. But we need to use energy wisely if we want to help slow global warming.

How? To do most of these things, you need to use electricity. Most electricity is generated in power plants that use coal, gas, and oil. Burning fossil fuels produces greenhouse gases. As an individual, you can conserve electricity, use public transportation, support alternative technologies that replace internal combustion engines, and produce some of your own food in a garden.

Oh, and gardening has a second benefit to the environment. Plants use CO_2 in photosynthesis. In the process, CO_2 is removed from the atmosphere. So grow plants, particularly trees. They take up a lot of CO_2 and hold it in the form of cellulose for a very long time.

Adapted from: The Environmental Protection Agency's (EPA) Global Warming Kids Page. www.epa.gov/globalwarming/kids/index.html

That's how Professor Tetsuya Theodore Fujita described the beginning of his lifelong fascination with severe weather. He was 27 years old when he made those first observations. After completing his degree in meteorology at the University of Tokyo, Fujita joined the faculty at the University of Chicago. Some of the most powerful storms in the world occur in the stretch of land that runs from Texas, north and east through Oklahoma and Kansas, and into the Midwest. This corridor, known as Tornado Alley, was where Professor Fujita focused much of his study over the next several decades.

Shortly after his arrival in Chicago, Ted, as he was popularly known, began analyzing individual thunderstorms in detail. He observed and recorded temperature, pressure, and wind data and related them to the development of the huge, dark clouds that form during thunderstorms. His interest in thunderstorms led directly to tornadoes. It was in the area of tornado study that Ted established his reputation as a pioneer researcher. Among his peers he was nicknamed Mr. Tornado, in recognition of his great contributions.

Ted introduced the concept of tornado families, a group of individual tornadoes, each with a unique path, spawned by the same massive thunderstorm. Prior to this, long damage paths were thought to be made

by a single tornado. Ted discovered that, as a thunderstorm advanced, two, three, or more funnels might form, touch down, dissipate, and re-form later to create more destruction on the ground. On occasion, two or more funnels might extend from the storm front at the same time. He introduced new concepts of thunderstorm architecture and developed terms like "wall cloud" and "tail cloud."

In the late 1960s, Ted's analysis of the Palm Sunday outbreak of 1965 changed the course of how we view a tornado outbreak. For the first time, he mapped the entire outbreak in terms of tornado families. While multiple-vortex tornadoes are well known today, he was the first to identify their existence, based on damage patterns.

In the 1970s, he again revolutionized tornado climatology by giving us a system that linked damage and wind speed. Previously, all tornadoes were counted as equals. Ted quantified their force on a five-level scale, with force 5 (F5) being the most powerful and potentially destructive storms. For the Super Outbreak of 1974, Ted was able to develop Fujita scale-intensity contour maps for the entire path of many of the 148 tornadoes that raged that year. After 25 years, we still use his ideas and terminology.

In the late 1970s, Ted turned his attention to weather-related aircraft disasters. He identified two phenomena that had not been described before, the **downburst** and the **microburst.** Before this, meteorologists had been confused by a bewildering array of gusty winds in and around thunderstorms. By the 1980s, the downburst and microburst were understood as separate winds. They were known to be caused by events such as dry air moving into a thunderstorm.

In Ted's later years, he applied his knowledge to hurricanes and typhoons, which are the kind of weather that he had originally focused on in Japan, where he was born.

Weather is fairly predictable most of the time. During the summer months in the San Francisco Bay area, you can expect foggy mornings and late afternoons, with the possibility of sunshine midday. In the southeast United States, summer days are often hot and humid. In the Midwest and East, winters are usually cold, cloudy, and snowy. These are the normal conditions that people come to expect where they live.

It's the change from normal that catches people's attention, whether they see it on TV or experience it for themselves. Tornadoes, thunderstorms, windstorms, hurricanes, and floods are examples of what is known as **severe weather.** Severe weather is weather out-of-the-ordinary. It usually causes dangerous conditions that can damage property and threaten lives.

FLOOD

Floods occur when water overflows the natural or artificial banks of a stream or other body of water. The water moves over normally dry land or accumulates in low-lying areas. Floods often happen with heavy rainfall over short periods of time or when a large quantity of snow melts quickly.

Floods may also occur behind ice dams on rivers, during very high tides, or following tsunamis (huge waves) caused by earthquakes under water. **Flash floods** are short, rapid, unexpected flows of muddy water rushing down a canyon. They are often caused by thunderstorms occurring over mountains, during which large quantities of water flow down a single canyon.

Johnstown, Pennsylvania, 1889

In 1889, a flood of disastrous proportions hit the city of Johnstown, Pennsylvania. An earthen dam collapsed after heavy rains. A great wall of water rushed down the Conemaugh River valley at speeds up to 64 kilometers (km) an hour. The 10-meter-high wall of water devastated the town, washing away most of the northern half of the city, killing 2209 people, and destroying 1600 homes.

DROUGHT

Droughts are less than normal precipitation over a long period of time. They usually cause water shortages, including low flow or no flow in streams. Soil moisture and groundwater levels decrease. Droughts are especially disastrous to agriculture.

The Dust Bowl, 1930s

During the early 1930s, an area of the United States that includes parts of Colorado, Kansas, New Mexico, and the panhandles of Texas and Oklahoma was named the Dust Bowl. A severe drought occurred after years of poor land management. The native grasses had been removed, exposing the topsoil. Strong spring winds blew away the topsoil, causing "black blizzards." Thousands of families left the area at the height of the Great Depression.

Pakistan, 2000

Large parts of Pakistan were hit by drought in 2000. No rain fell in the Baluchistan area for more than 3 years. Many people were forced to leave some of the more remote villages. They migrated to cities in search of food and water. This required a major support effort by the Pakistani government and several international relief agencies.

HAIL

Hail is frozen precipitation in the form of balls of ice. The diameter of hailstones ranges from 0.5 to 10 centimeters (cm). Hail usually forms during thunderstorms when strong updrafts (vertical winds) move through cumulonimbus clouds in which temperatures are near or below freezing.

Northwest Missouri, September 5, 1898

A huge hailstorm assaulted Nodaway County in northwest Missouri on September 5, 1898. The hail remained on the ground for 52 days and left the fields unworkable for 2 weeks. On October 27, there was still enough hail left in ravines for the local residents to make ice cream.

Selden, Kansas, June 3, 1959

On June 3, 1959, a severe hailstorm struck the town of Selden in northwest Kansas. Hail fell for 85 minutes. The storm covered an area 10 km by 14.4 km with hailstones to a depth of 46 cm. The storm caused $500,000 worth of damage. That was a lot of money in 1959.

HURRICANE

Hurricanes are cyclones, moving wind systems that rotate around an eye, or center of low atmospheric pressure. Hurricanes form over warm tropical seas. Wind speeds are more than 64 knots (119 km/h) in a hurricane. The term *hurricane* is used for Northern Hemisphere cyclones east of the International Dateline to the Greenwich Meridian. The term **typhoon** is used for Pacific cyclones north of the equator west of the International Dateline. Hurricanes produce dangerous winds, heavy rains, and flooding. They can cause great property damage and loss of life in coastal areas.

Galveston, Texas, September 8, 1900

The hurricane that struck Galveston, Texas, on September 8, 1900, is considered one of the worst natural disasters in U.S. history. More than 6000 men, women, and children lost their lives during the Great Storm. It is estimated that the winds reached 250 to 333 km/h. The tidal surge was probably 4.6 to 6.2 m. The highest elevation on Galveston Island in 1900 was 2.7 m. More than 3600 homes were destroyed, with whole blocks of homes totally wiped out, leaving only a few bricks behind.

Southern Florida, August 1992

Hurricane Andrew was a relatively small, but ferocious, hurricane that devastated areas in the northwest Bahamas, southern Florida, and south central Louisiana during August 1992. Hurricane Andrew was the most expensive natural disaster in U.S. history at that time, causing nearly $25 billion worth of damage. Forty lives were lost, and almost 250,000 people were left homeless.

LIGHTNING AND THUNDER

Lightning is a visible electric discharge produced by thunderstorms. Not everyone agrees why lightning happens, but what happens is pretty well understood. Lightning travels from a cloud to the ground. The electric charge moves downward in approximately 46-m steps called **step leaders.** This flow continues until the charge reaches something on the ground that is a good conductor. The circuit is completed.

The whole event usually takes less than half a second, and a lightning bolt can reach 200 million volts.

Since 1989, U.S. meteorologists have been able to detect lightning with a network of antennas. An average of 20 million cloud-to-ground strikes happen every year over the continental United States.

Thunder is the explosive sound that usually accompanies lightning. The sound is caused by the rapidly expanding gases in the atmosphere along the path of the lightning. How loud and what type of sound you hear depends on atmospheric conditions and how far away you are from the flash.

Perth, Australia, December 16, 1998

An apprentice jockey, Damion Beckett, and champion gelding Brave Buck died instantly when lightning struck them. They were training at Ascot racecourse in Perth about 5 a.m. on December 16, 1998.

Durham, North Carolina, July 7, 1995

Lightning during an early-evening thunderstorm killed a golfer. The man took cover from the storm in a wooden shed at a golf course. Witnesses say the lightning first struck a tree and then bounced to the shed. Other golfers sustained scrapes and burns from the accident. All were dazed and in shock. Later they said that the lightning was the brightest they had ever seen and that the thunder came immediately.

TORNADO

Tornadoes are rapidly rotating columns of air that extend from a thunderstorm to the ground. Wind speeds in a tornado can reach 417 km/h or more. The path of a tornado can be more than 1 km wide and 83 km long.

Xenia, Ohio, April 3, 1974

On April 3 and 4, 1974, a super tornado outbreak struck the states of Alabama, Georgia, Illinois, Indiana, Kentucky, Michigan, Mississippi, North Carolina, Ohio, South Carolina, Tennessee, Virginia, and West Virginia. Especially devastating tornadoes struck Ohio during the afternoon and early evening of April 3. The town of Xenia was hardest hit. Thirty people died, more than 1100 were injured, and more than 1000 homes were destroyed. The path of damage ranged from 0.4 to 0.83 km wide.

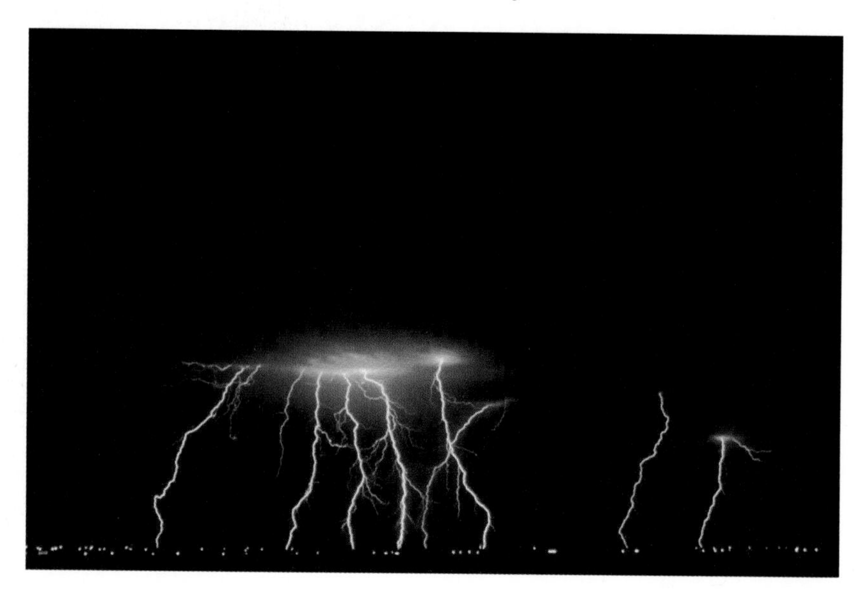

Oklahoma City, Oklahoma, May 3, 1999

Eight supercell thunderstorms produced at least 59 different tornadoes in central Oklahoma alone on May 3, 1999. Many of these tornadoes were very violent and long-lasting. They made direct hits on several populated areas, including Oklahoma City. At least 40 people died in Oklahoma because of the twisters, and 675 were injured. Total damage was as much as $1.2 billion. Tornadoes also caused extensive damage to the Wichita, Kansas, metro area.

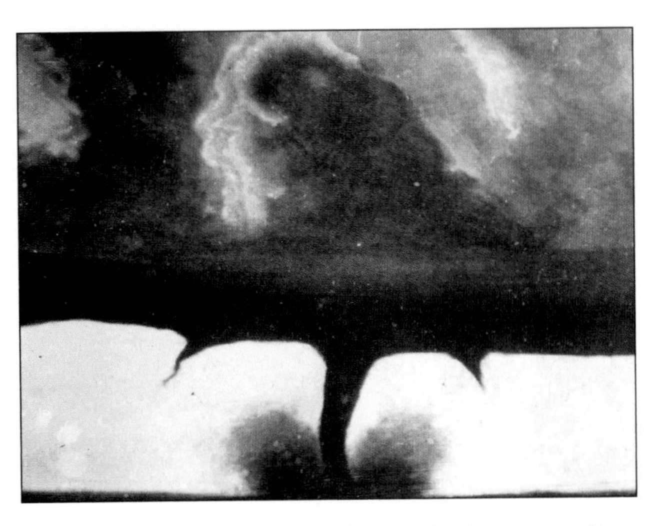

This is the oldest known photo of a tornado. It was taken 35 km southwest of Howard, South Dakota, on August 28, 1884.

Salt Lake City, Utah, August 11, 1999

The largest tornado ever to strike Salt Lake City happened on August 11, 1999. Tornadoes are rare in Utah. Winds from this tornado reached 188 to 262 km/h. One person died, 99 people were hurt, and 387 homes were damaged. Damage estimates reached $3.2 million.

THUNDERSTORM

Thunderstorms produce rapidly rising air currents, usually resulting in heavy rain or hail along with thunder and lightning. A thunderstorm is classified as severe when it produces one or more of the following:

- Hail 1.9 cm or greater in diameter
- Winds gusting in excess of 96 km/h
- A tornado

West Central Texas, October 22, 1996

On October 22, 1996, surface temperatures dropped from 10°C to around 0°C in an area of west central Texas. Bands of convection developed. The storm increased in intensity very quickly. Dime- to egg-sized hail fell, as well as large amounts of sleet and light snow.

Fort Worth, Texas, May 5, 1995

An isolated severe storm became the costliest thunderstorm in U.S. history when it devastated the area in and around Fort Worth, Texas, on May 5, 1995. More than 100 people were injured, mostly by softball-sized hail that pelted people attending an outdoor Mayfest. Winds reached 100 to 117 km/h (60 to 70 mph), driving hail the size of grapefruit in some areas. The large hail and high winds damaged hundreds of homes, businesses, and vehicles. Damage totals reached more than $2 billion.

BLIZZARD

Blizzards are severe storms with low temperatures, strong winds, and large quantities of snow. Blizzards have winds of more than 51 km/h and enough snow to limit visibility to 150 m or less. A severe blizzard has winds of more than 72 kph, near zero visibility, and temperatures of –12°C or lower.

East Coast, United States, February 1–14, 1899

A cold wave hit the East Coast during February 1–14, 1899, causing a huge blizzard and bitter-cold temperatures from the Rockies to the Atlantic Ocean. Snow fell from Louisiana to Georgia and extended northeast into New England. Florida experienced its first blizzard. Temperatures fell below freezing there for the first time. Up to 61 cm of snow fell across much of the Northeast. Winds gusted to 58 km/h. Jacksonville, Florida, measured its greatest snowfall ever at just over 4 cm. Tampa received its first measurable snow on record.

Chicago, January 26–27, 1967

The city of Chicago set records on January 26 and 27 with the worst blizzard ever recorded there. More than 52 cm of snow fell in 29 hours and 8 minutes. Trains froze and couldn't move. Buses were stranded. O'Hare Airport was closed for 3 days.

Washington, DC, February 22, 1979

Known as the Presidents' Day Storm, this blizzard covered much of the Northeast, closing down Washington, DC. Ice formed on the Potomac. Some mid-Atlantic areas received more than 44 cm of snow. Observers in Baltimore reported snow falling at a rate of 11 cm per hour.

Super Storm, March 12–15, 1993

Twenty-six states were affected by a low-pressure system during March 12–15, 1993. The system brought heavy snow and strong winds. Birmingham, Alabama, received 28.6 cm of snow, while Chattanooga, Tennessee, measured 44 cm by the time the storm passed. Winds gusted up to 83 km/h as temperatures fell rapidly. Parts of Long Island, New York, recorded wind gusts of 155 km/h. Over 50 tornadoes hit Florida. The storm took more than 200 lives.

WINDSTORM

Sometimes strong winds that are not directly connected with thunderstorms, tornadoes, or hurricanes occur. **Straight-line winds** are strong winds that have no rotation. The winds can travel at speeds of more than 167 km/h. If no one is around to observe what happens, it can be difficult to tell if damage came from a straight-line wind or a tornado.

Microbursts are small, very intense downdrafts. They affect areas less than 4 km wide. Microbursts can produce winds of more than 280 km/h. They typically last less than 10 minutes. They are often associated with thunderstorms. Microbursts often cause significant ground damage and are a threat to aviation.

Dust devils are small rotating winds not associated with a thunderstorm. They become visible when they collect dust or debris. Dust devils form when air is heated by a hot surface during fair, hot weather. You are most likely to see a dust devil in arid or semiarid regions.

Dust storms are rare conditions in which strong winds carry dust over a large area. They occur when there are drought conditions. In desert areas, sandstorms occur.

China, April 1998

Parts of China and Mongolia experienced a major dust storm in April 1998. Twelve people were reported missing after the storm, whose winds reached gale force. The storm blanketed ten cities and districts. Power and water supplies were cut. A trace of yellow dust carried by the winds was deposited on nearby deserts. Some of the dust was blown all the way to the United States.

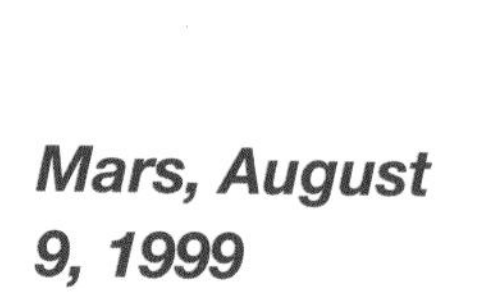

Mars, August 9, 1999

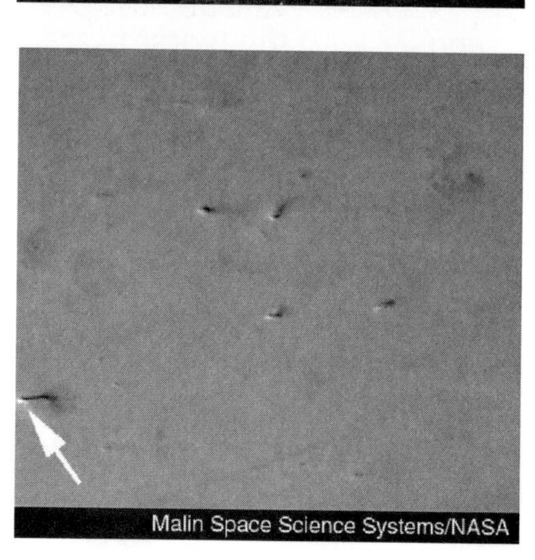

The Mars Global Surveyor *photographed a dust storm blowing across Mars's northern plains. A Martian dust storm can cover most of the planet. Scientists have also identified dust devils on Mars photographs. This dust devil has a diameter of 8 km.*

For more information about severe weather, visit this National Weather Service website: www.nws.noaa.gov/om/severeweather/index.shtml

DATA
Table of Contents

Humidity Calculator

Dry-bulb temp. (°C)	Relative humidity (%)																			
	Difference between wet- and dry-bulb temperature (°C)																			
	1°	2°	3°	4°	5°	6°	7°	8°	9°	10°	11°	12°	13°	14°	15°	16°	17°	18°	19°	20°
–10	67	35																		
–9	69	39	9																	
–8	71	43	15																	
–7	73	48	20																	
–6	74	49	25																	
–5	76	52	29	7																
–4	77	55	33	12																
–3	78	57	37	17																
–2	79	60	40	22																
–1	81	62	43	26	8															
0	81	64	46	29	13															
1	83	66	49	33	17															
2	84	68	52	37	22	7														
3	84	70	55	40	26	12														
4	85	71	57	43	29	16														
5	86	72	58	45	33	20	7													
6	86	73	60	48	35	24	11													
7	87	74	62	50	38	26	15													
8	87	75	63	51	40	29	19	8												
9	88	76	64	53	42	32	22	12												
10	88	77	66	55	44	34	24	15	6											
11	89	78	67	56	46	36	27	18	9											
12	89	78	68	58	48	39	29	21	12											
13	89	79	69	59	50	41	32	23	15	7										
14	90	79	70	60	51	42	34	26	18	10										
15	90	80	71	61	53	44	36	27	20	13	6									
16	90	81	71	63	54	46	38	30	23	15	8									
17	90	81	72	64	55	47	40	32	25	18	11									
18	91	82	73	65	57	49	41	34	27	20	14	7								
19	91	82	74	65	58	50	43	36	29	22	16	10								
20	91	83	74	66	59	51	44	37	31	24	18	12	6							
21	91	83	75	67	60	53	46	39	32	26	20	14	9							
22	92	83	76	68	61	54	47	40	34	28	22	17	11	6						
23	92	84	76	69	62	55	48	42	36	30	24	19	13	8						
24	92	84	77	69	62	56	49	43	37	31	26	20	15	10	5					
25	92	84	77	70	63	57	50	44	39	33	28	22	17	12	8					
26	92	85	78	71	64	58	51	45	40	34	29	24	19	14	10	5				
27	92	85	78	71	65	58	52	47	41	36	31	26	21	16	12	7				
28	93	85	78	72	65	59	53	48	42	37	32	27	22	18	13	9	5			
29	93	86	79	72	66	60	54	49	43	38	33	28	24	19	15	11	7			
30	93	86	79	73	67	61	55	50	44	39	35	30	25	21	17	13	9	5		
31	93	86	80	73	67	61	56	51	45	40	36	31	27	22	18	14	11	7		
32	93	86	80	74	68	62	57	51	46	41	37	32	28	24	20	16	12	9	5	
33	93	87	80	74	68	63	57	52	47	42	38	33	29	25	21	17	14	10	7	
34	93	87	81	75	69	63	58	53	48	43	39	35	30	28	23	19	15	12	8	5
35	93	87	81	75	69	64	59	54	49	44	40	36	32	28	24	20	17	13	10	7

RAINDROPS AND CLOUD DROPLETS

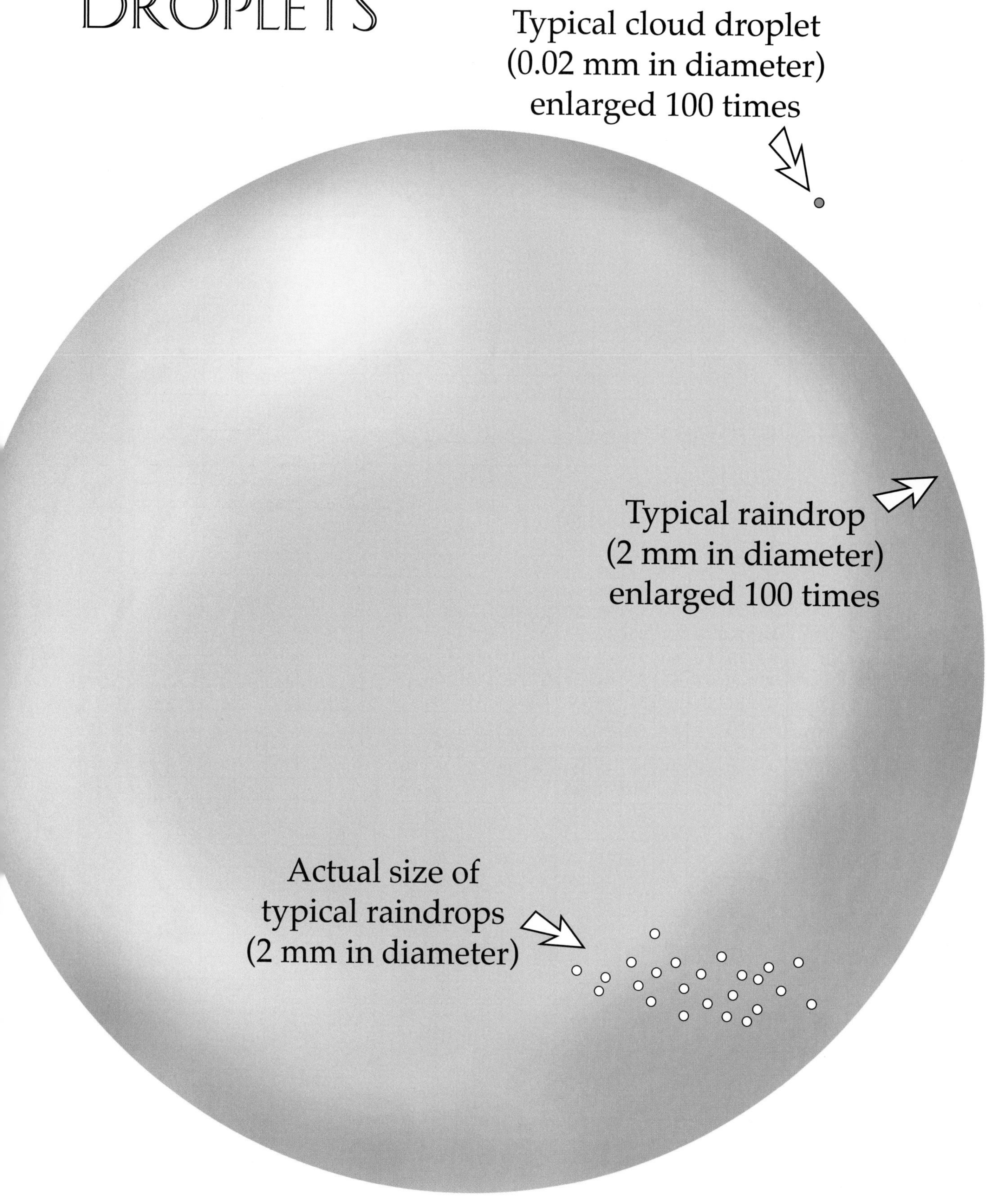

Typical cloud droplet
(0.02 mm in diameter)
enlarged 100 times

Typical raindrop
(2 mm in diameter)
enlarged 100 times

Actual size of
typical raindrops
(2 mm in diameter)

Weather-Balloon Sounding Data
May 9, 2000

Phoenix, Arizona

Altitude (m)	Air temp. (°C)	Dew point (°C)
75	33	8
300	31	6
550	28	5
800	26	5
1000	24	5
1250	21	4
1500	19	3
1750	17	2
2000	15	0
3000	9	–14
4400	4	–22
5800	–7	–26
8500	–28	–43
12,000	–59	–66

Chicago, Illinois

Altitude (m)	Air temp. (°C)	Dew point (°C)
40	16	14
250	14	13
500	13	12
700	13	12
900	12	12
1200	12	11
1400	11	10
1900	9	8
2400	6	5
3000	3	3
3600	0	–2
5000	–9	–9
8300	–30	–32
12,000	–60	–68

Boston, Massachusetts

Altitude (m)	Air temp. (°C)	Dew point (°C)
16	10	9
150	9	5
600	8	6
800	9	9
1200	9	9
1500	9	9
1700	9	9
2100	8	7
4000	–4	–5
4900	–8	–9
5300	–10	–12
6100	–17	–17
7000	–22	–24
9100	–38	–42

San Francisco, California

Altitude (m)	Air temp. (°C)	Dew point (°C)
10	12	9
170	12	9
400	10	9
600	9	7
800	9	4
1000	10	0
1300	10	–3
1500	9	–3
2000	8	–2
2600	6	–3
5000	–6	–21
7500	–22	–33
9500	–38	–44
12,000	–60	–64

Water-Cycle Game Rules: Plain Version

WHAT YOU ROLL	WHAT HAPPENS TO YOU	WHERE YOU GO
GROUNDWATER		
1	Water filters into a river.	River
2 or 3	Water filters into a lake.	Lake
4, 5, or 6	Water stays underground in an aquifer. Roll again.	Groundwater
RIVER		
1	Water flows into a lake.	Lake
2	Water filters into the soil.	Soil
3	Water flows into the ocean.	Ocean
4	An animal drinks water.	Animal
5	Water heats up and evaporates.	Atmosphere
6	Water remains in the river. Roll again.	River
ANIMAL		
1 or 2	Water is excreted through feces and urine.	Soil
3, 4, or 5	Water is respired or evaporated from the body.	Atmosphere
6	Water is incorporated into the body. Roll again.	Animal
SOIL		
1	Water is absorbed by plant roots.	Plant
2	Soil is saturated, so water runs into a river.	River
3	Water filters into the soil.	Soil
4 or 5	Heat evaporates the water.	Atmosphere
6	Water remains on the surface, in a puddle, or on a soil particle. Roll again.	Soil
ATMOSPHERE		
1	Water condenses and falls on soil.	Soil
2	Water condenses and falls as snow on a glacier.	Glacier
3	Water condenses and falls on a lake.	Lake
4 or 5	Water condenses and falls on an ocean.	Ocean
6	Water remains as vapor in the atmosphere. Roll again.	Atmosphere
GLACIER		
1	Ice melts and water filters into the ground.	Groundwater
2	Ice sublimates (turns directly from ice into water vapor) and goes into the atmosphere.	Atmosphere
3	Ice melts and water flows into a river.	River
4	Ice melts and water flows into the ocean.	Ocean
5 or 6	Ice stays frozen in the glacier. Roll again.	Glacier
LAKE		
1	Water filters into the soil.	Soil
2	An animal drinks water.	Animal
3	Water flows into a river.	River
4	Water heats up and evaporates.	Atmosphere
5 or 6	Water remains within a lake or estuary. Roll again.	Lake
OCEAN		
1 or 2	Water heats up and evaporates.	Atmosphere
3, 4, 5, or 6	Water remains in the ocean. Roll again.	Ocean
PLANT		
1, 2, 3, or 4	Water leaves a plant through the process of transpiration.	Atmosphere
5 or 6	Water is used by a plant and stays in cells. Roll again.	Plant

Water-Cycle Game Rules: Global-Warming Version

WHAT YOU ROLL	WHAT HAPPENS TO YOU	WHERE YOU GO
GROUNDWATER		
1 or 2	Water filters into a river.	River
3, 4, 5, or 6	Water filters into a lake.	Lake
7, 8, 9, 10, 11, or 12	Water stays underground in an aquifer. Roll again.	Groundwater
RIVER		
1 or 2	Water flows into a lake.	Lake
3 or 4	Water filters into the soil.	Soil
5 or 6	Water flows into the ocean.	Ocean
7 or 8	An animal drinks water.	Animal
9 or 10	Water heats up and evaporates.	Atmosphere
11 or 12	Water remains in the river. Roll again.	River
ANIMAL		
1, 2, 3, or 4	Water is excreted through feces and urine.	Soil
5, 6, 7, 8, 9, or 10	Water is respired or evaporated from the body.	Atmosphere
11 or 12	Water is incorporated into the body. Roll again.	Animal
SOIL		
1 or 2	Water is absorbed by plant roots.	Plant
3 or 4	Soil is saturated, so water runs into a river.	River
5 or 6	Water filters into the soil.	Soil
7, 8, 9, or 10	Water heats up and evaporates.	Atmosphere
11 or 12	Water remains on the surface, in a puddle, or on a soil particle. Roll again.	Soil
ATMOSPHERE		
1 or 2	Water condenses and falls on soil.	Soil
3 or 4	Water condenses and falls as snow on a glacier.	Glacier
5 or 6	Water condenses and falls on a lake.	Lake
7, 8, 9, or 10	Water condenses and falls on an ocean.	Ocean
11 or 12	Water remains as vapor in the atmosphere. Roll again.	Atmosphere
GLACIER		
1 or 2	Ice melts and water filters into the ground.	Groundwater
3, 4, or 5	Ice sublimates (turns directly from ice into water vapor) and goes into the atmosphere.	Atmosphere
6, 7, or 8	Ice melts and water flows into a river.	River
9, 10, or 11	Ice melts and water flows into the ocean.	Ocean
12	Ice stays frozen in the glacier. Roll again.	Glacier
LAKE		
1 or 2	Water filters into the soil.	Soil
3 or 4	An animal drinks water.	Animal
5 or 6	Water flows into a river.	River
7, 8, or 9	Water heats up and evaporates.	Atmosphere
10, 11, or 12	Water remains within a lake or estuary. Roll again.	Lake
OCEAN		
1, 2, 3, 4, or 5	Water heats up and evaporates.	Atmosphere
6, 7, 8, 9, 10, 11, or 12	Water remains in the ocean. Roll again.	Ocean
PLANT		
1, 2, 3, 4, 5, 6, 7, 8, or 9	Water leaves a plant through the process of transpiration.	Atmosphere
10, 11, or 12	Water is used by a plant and stays in cells. Roll again.	Plant

NORTH AMERICAN AIR MASSES

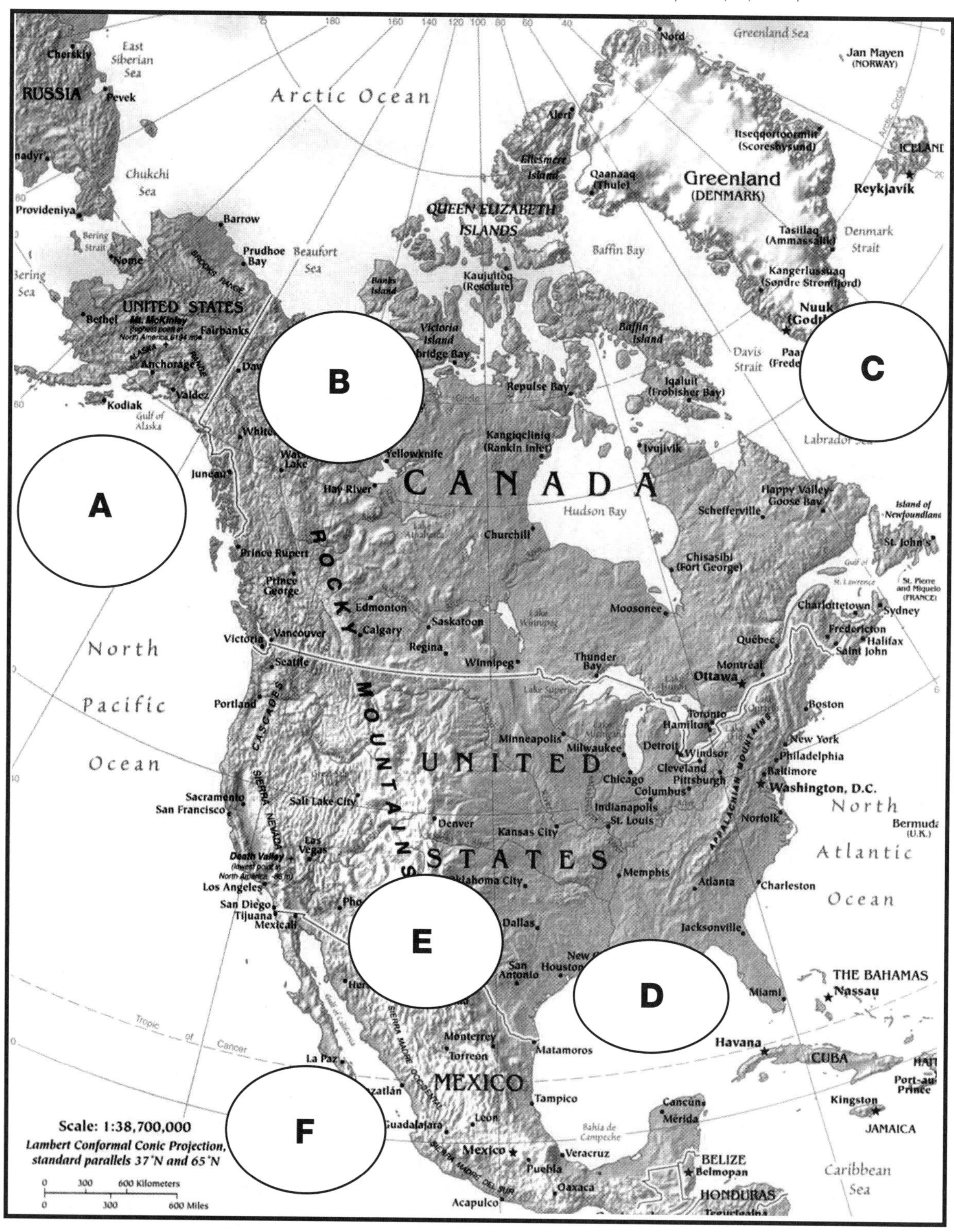

Continental (c): **related to the land**
Maritime (m): **related to the ocean or sea**
Polar (P): **areas near the poles, including Canada and the Arctic Ocean**
Tropical (T): **areas in the tropics, Gulf of Mexico, Mexico, and southwestern United States**

FRONTS

COLD FRONT

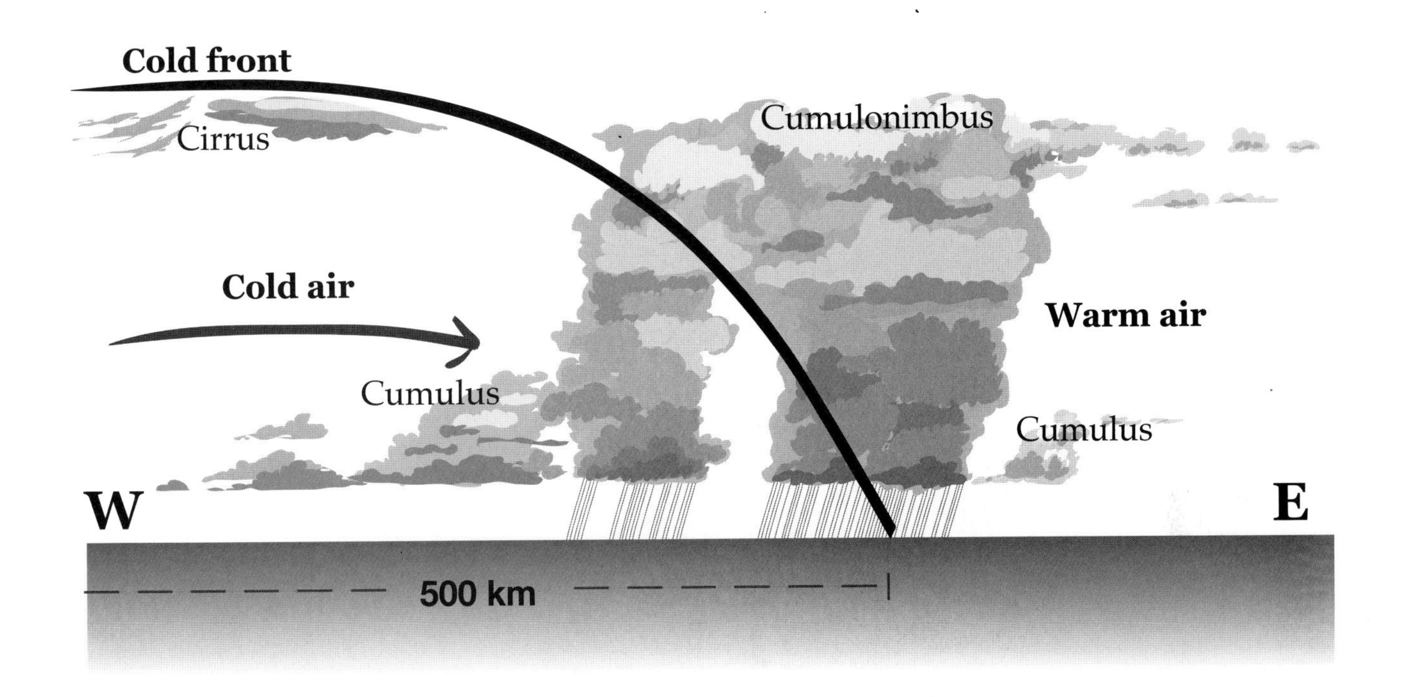

WARM FRONT

Warm front

Warm air

Cold air

W

E

1000 km

WEATHER AND FRONTS

COLD FRONT

WEATHER OBSERVATION	BEFORE FRONT PASSES	WHILE FRONT PASSES	AFTER FRONT PASSES
Winds	South to southwest	Gusty, shifting	West to northwest
Temperature	Warm	Sudden drop	Drops steadily
Pressure	Falls steadily	Small drop, then sharp rise	Rises steadily
Clouds	Cloud cover increases: cirrus, cirrostratus, cumulonimbus	Cumulonimbus	Cumulus
Precipitation	Short period of showers	Heavy rain, sometimes with thunderstorms, including hail	Showers followed by clearing
Visibility	Fair to poor, hazy	Poor, then improving	Good, unless there are showers
Dew point	High and steady	Drops sharply	Continues to lower

WARM FRONT

WEATHER OBSERVATION	BEFORE FRONT PASSES	WHILE FRONT PASSES	AFTER FRONT PASSES
Winds	South to southeast	Variable	South to southwest
Temperature	Cool/cold, slowly warming	Steadily rising	Warming, then steady
Pressure	Usually falling	Leveling off	Rises slightly, then falls
Clouds	Clouds usually in this order: cirrus, cumulostratus, altostratus, nimbostratus, stratus, then fog; sometimes cumulonimbus also possible in summer	Stratus and variations of stratus	Clearing followed by stratocumulus; sometimes cumulonimbus appear in summer
Precipitation	Light to moderate rain or drizzle	Drizzle or none	Usually none, but sometimes light rain or showers
Visibility	Poor	Poor, but improving	Hazy or fair
Dew point	Steady rise	Steady	Rising, then steady

Sample Weather-Map Symbol

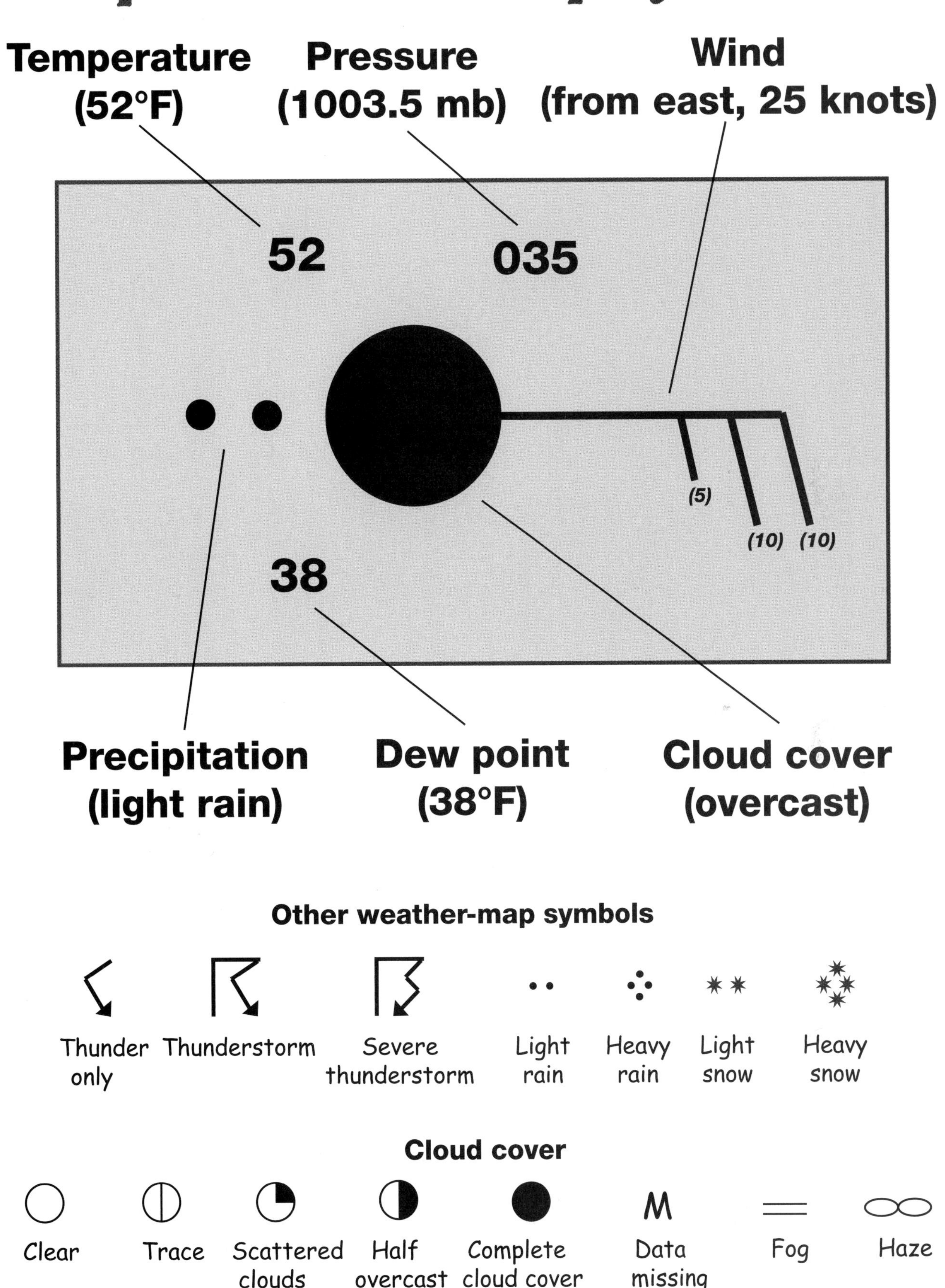

Surface Observations
October 18, 2000

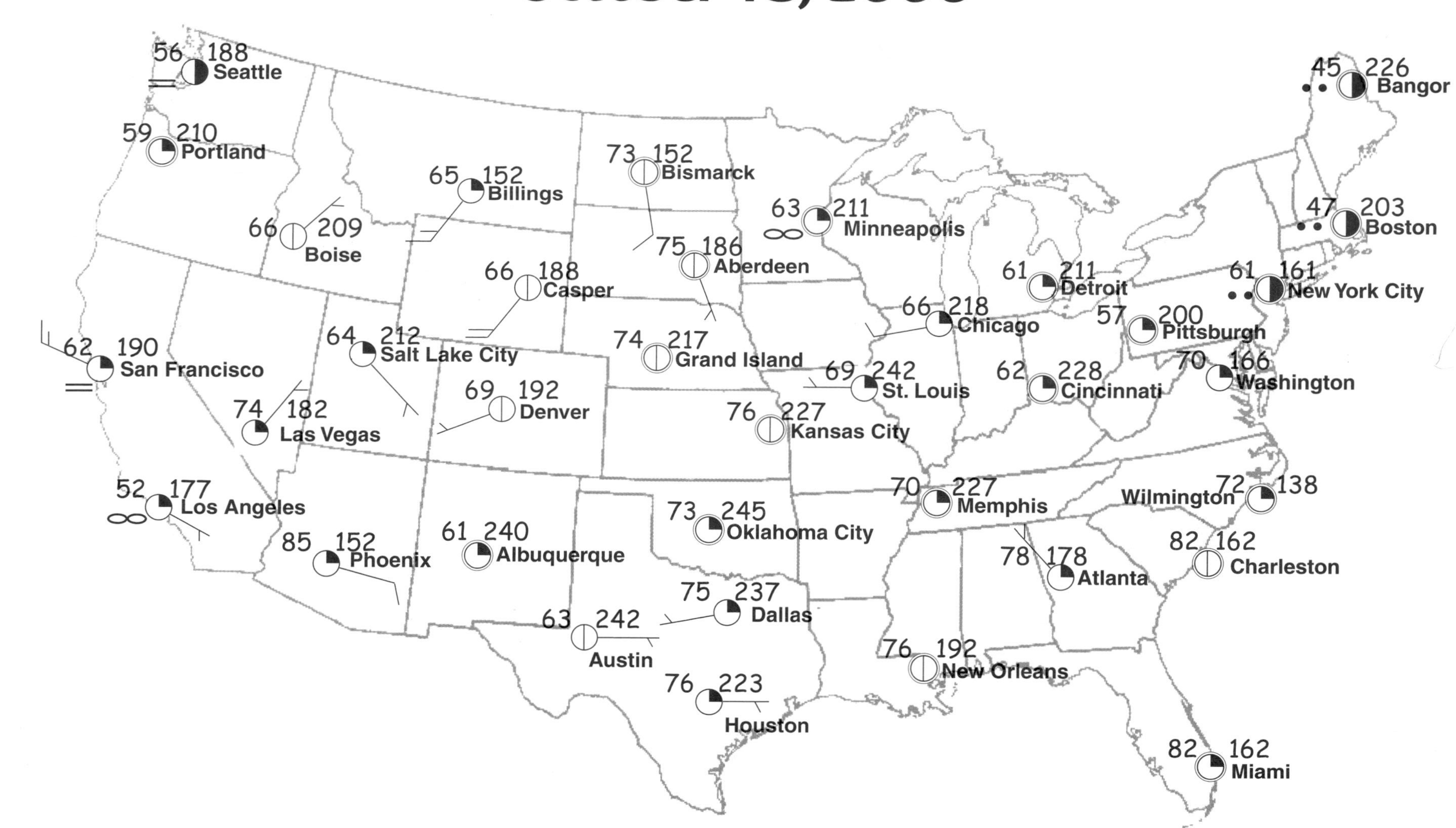

GOES-8 WEATHER-SATELLITE IMAGE

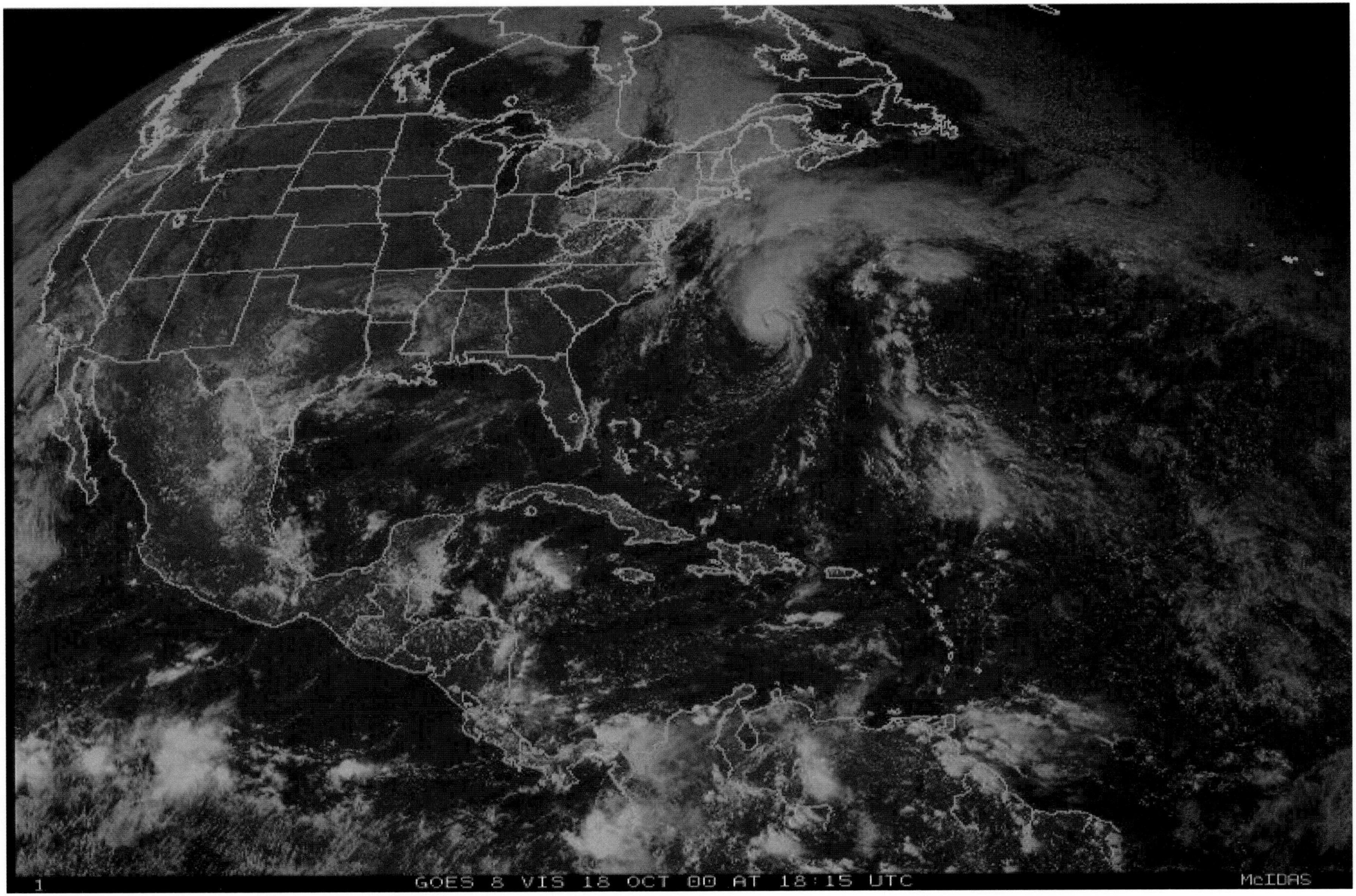

GOES 8 VIS 18 OCT 00 AT 18:15 UTC

McIDAS

OCTOBER 18, 2000 2345Z www.goes.noaa.gov

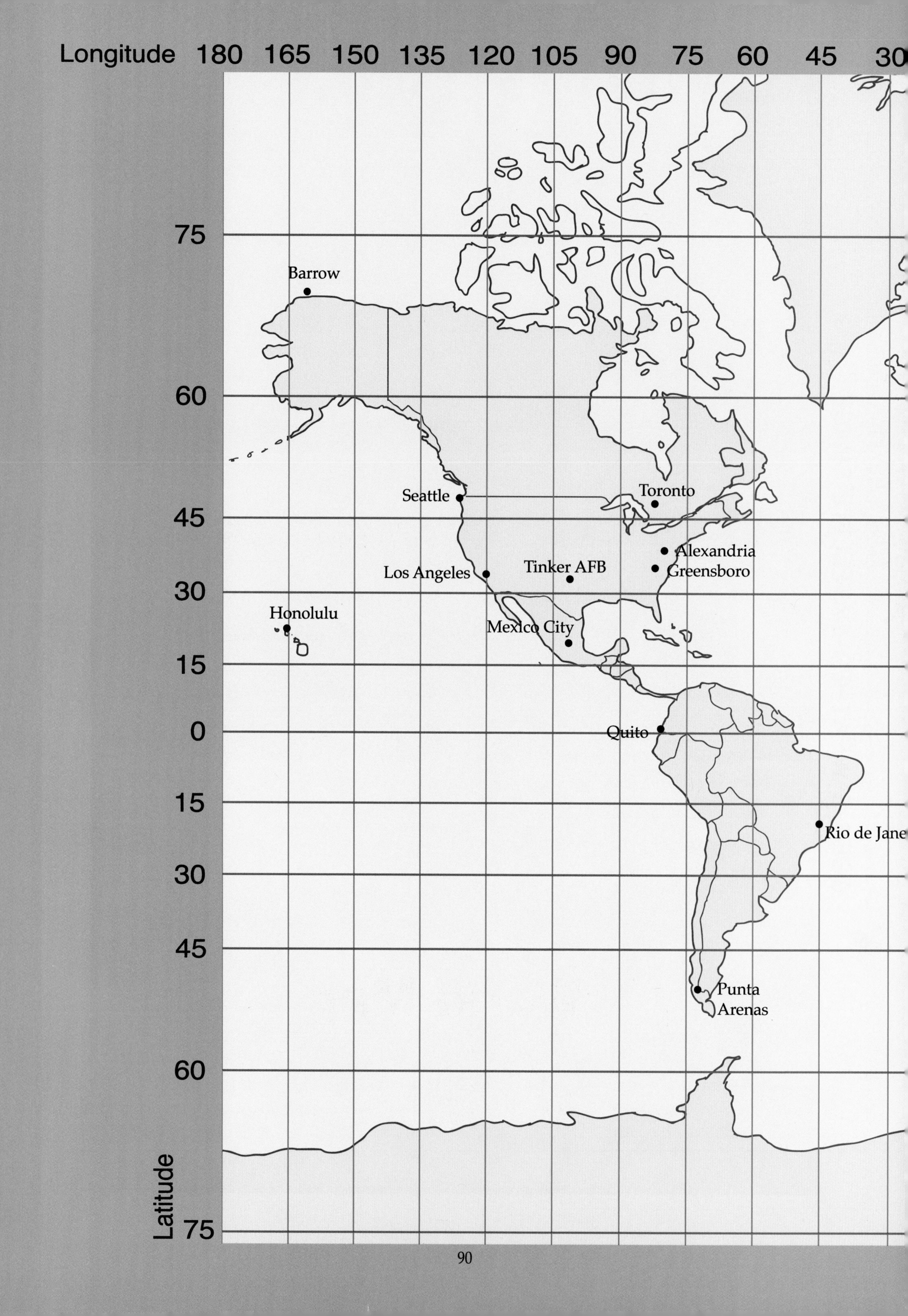
Longitude
180
165
150
135
120
105
90
75
60
45
30
75
60
45
30
15
0
15
30
45
60
75
Latitude
Barrow
Seattle
Toronto
Alexandria
Greensboro
Tinker AFB
Los Angeles
Honolulu
Mexico City
Quito
Rio de Jane
Punta
Arenas

0
15
30
45
60
75
90
105
120
135
150
165
180
Stockholm
Moscow
Paris
Rome
Casablanca
Cairo
New Delhi
Hong Kong
Sendai
Nairobi
Johannesburg
Sydney
Auckland
World Map